Space and Time Redux

Also, by this Author

Space, Time, and Deity Redux
History (and other things) Reconsidered
Consciousness and Deity Reconsidered

Space and Time Redux

Proposing New Notions About
*Space and Time
for the 21st Century*

John H. Silver

Beagle Publishing 1-7A

ISBN-13: 9781691484164

Cover photo: The distinctive blue bubble appearing to en-circle WR 31a, and its uncatalogued stellar sidekick, is a Wolf–Rayet nebula — an interstellar cloud of dust, hy-drogen, helium, and other gases. Credit: ESA/Hubble & NASA. Acknowledgement: Judy Schmidt

Beagle Publishing

Contents

Introduction

This book is primarily the result of two incidents that occurred in the early 2000s. The first was a severe personal tragedy that compelled me to seek answers concerning life and purpose, and the second was the fortuitous encounter with the English philosopher Samuel Alexander's book *Space, Time, and Deity (1920)*[1] that is a two-volume compilation of his Gifford Lectures of 1916 - 1918.[2] This encounter was fortuitous because since the first incident, as mentioned, I had been searching for some answers to questions about the universe, life, and all that – but like so many others I found that the traditional answers did not fully satisfy. That is, until I read Alexander's lectures on space, time, and deity.

---

[1] Samuel Alexander (1920), *Space, Time, and Deity*. The Gifford Lectures at Glasgow. https://archive.org/details/spacetime-anddeit00alexuoft>.

[2] These lectures have been delivered annually since 1888 and here is what the Gifford Lectures website says: "The prestigious Gifford Lectureships were established by Adam Lord Gifford (1820–1887), a senator of the College of Justice in Scotland. The purpose of Lord Gifford's bequest to the universities of Edinburgh, Glasgow, St. Andrews and Aberdeen was to sponsor lectures to 'promote and diffuse the study of Natural Theology in the widest sense of the term—in other words, the knowledge of God'." Since the first lecture in 1888, Gifford Lecturers have been recognized as pre-eminent thinkers in their respective fields [https://www.giffordlectures.org/].

Alexander's ideas about space and time were formed shortly after the first great 20th century revolution in science - the theories of relativity - were constructed, but not yet completely understood, and before the second great revolution in science – quantum theory – was constructed. However, perhaps of equal importunacy, in Alexander's time the universe was thought to be very small compared to our current knowledge. Even though somewhat outdated, Alexander's lectures were important to me as a starting point that jump-started my ideas about space and time that eventually led to this book.

To be sure, cosmology and other related observational disciplines have many questionable notions, however, enough is firmly established that we can state with confidence that new knowledge about the vastness of the observable universe suggests, or even requires, that we reconsider our notions about space and time.

*   *   *

This book is an abridgement of an earlier work - *Space, Time, and Deity Redux* and consists selected sections of that earlier book including all of part 1 – Space and Time, along with selected elements from part 2 – Perception and Reality.[3]

The primary reason for this splitting of *Space, Time, and Deity Redux* into two volumes was that

---

[3] Also, as a complementary volume, the third part – Consciousness and Deity – from the earlier work has also been split out as a separate book (*Consciousness and Deity Reconsidered*).

I believe that there are readers who might be interested in one or the other – but not both topics. That is, some might prefer the space and time discussions without the consciousness and deity part while others would prefer the other way around.

On a practical side, this abridgement of the earlier work provided the opportunity to engage in corrective editing and other housekeeping.

*   *   *

Until about a hundred years ago, the universe was considered to be an expanse of only a few thousand light years and was composed only of what we now call the milky way. Today, the known size of the observable universe has increased about a million-fold and is now measured in billions of light years.

Current knowledge of the vastness of the universe and the unexpected strangeness of the micro-world require that we rethink some of our most strongly held beliefs. We need to remember that many of our beliefs originated thousands of years ago, in the setting of a small universe, and where earth was the center of that universe. Although modern cosmology has many questionable notions, enough of it is established to the point where rethinking our place in the universe is both desirable and necessary.

In some ways, conceptualizing time has always been more challenging and problematic than space. In the case of space, we have sensual perception to assist us. We may not see space, but we do see that there is a separation between objects, both the separation of small objects nearby and the separation of heavenly bodies. This separation, the gaps between visible objects, is what we commonly call space.

On the other hand, time, a thing not directly perceived in any physical way by our sense perception, is an abstract mental construction that we use to explain the separation of events; in a manner, somewhat analogous to the way we use space to explain the separation of objects. The obvious difference is that the separation of objects is directly observed, while our sensation of time is not directly observed and is variable to the extreme. The only way we can visualize time is geometrically as a series of unequal linear durations such as an unequal dashed line (- - —— - - —).

To assist the reader in understanding something about the nature of these speculations, brief discussions about past and present scientific theories are presented in as non-technical manner as possible, while still providing some understanding of the science that has been developed over the last century or so.

As this work progresses, it becomes increasingly conceptual and speculative. In this way, I have attempted to open new notions about our world, which provides enrichment and a greater appreciation for the complexity and wonder of our world.

It is safe to say, that some things that now seem complex will be simplified and some things that seem simple will become more complex. Portions of this book may be common knowledge to some, new and novel to some, and, possibly, completely absurd to some [hopefully that not too many will have the latter experience]. Of course, the same reader could have all three experiences.

As a last word, I would like to thank my wife for her patience during the many hours I spent sequestered while working on this project.

Santa Rosa, California
December 2019
John H. Silver
jhslbc@gmail.com
www.Things-Reconsidered.com

"In every investigation, in every extension of knowledge, we're involved in action. And in every action, we're involved in choice. And in every choice, we're involved in a kind of loss, the loss of what we didn't do. We find this in the simplest situation. ... Meaning is always obtained at the cost of leaving things out... In practical terms this means, of course, that our knowledge is always finite and never all encompassing. ... This makes ours an open world, a world without end."[4]

"The Red Queen shook her head. "You may call it 'nonsense' if you like," she said, "but I've heard nonsense, compared with which that would be as sensible as a dictionary!'"[5]

"About the nature of the external world we have at the outset nothing but hypotheses. Before we test them in any very exact way, we may with safety try to understand them. Perhaps what seemed the wildest of them all may turn out to be the very best."[6]

"Talent hits the target that others miss, genius hit targets others cannot see."[7]

---

4 J. R. Oppenheimer, "The Oppenheimer Years," Los Alamos Science Winter/Spring (1983) http://la-science.lanl.gov/lascience07.shtml, 18.

[5] Lewis Carroll (1871), *Through the Looking-Glass, and What Alice Found There* (New York: Barnes and Noble, 2012), 133.

[6] Josiah Royce *Mind and Reality* (1882), https://archive.org/details/mindreality00roycrich, 6.

[7] Ancient Chinese Proverb.

Considering Time and Space

Ask someone *'What time is it?'* and the response will be without hesitation. However, ask *'What is time?'* and you will get a very different response. That is, you are more likely to get a blank stare than you are an answer. Just to be sure, I tried this question on several unsuspecting victims, and received reactions ranging from suspecting a trick, to an incredulous look, and even a somewhat hostile reaction. After a quick explanation that the question is not a trick or a challenge, the unsuspecting victims usually relaxed.

Anyway, here are two examples of the more creative answers to the 'What is time?' question:

"I think I would say that time is the dimension required for there to be change and to measure change. And I think it is ultimately incremental, not continuous."

"...perceived as a continuum that passes in our lives, time is a static dimension that life passes through."

Although very different answers, both have a similarity in that each managed to avoid the circularity of using the word *time* to define time. Both responders also used the word *dimension*, but they disagreed on whether time is a continuum (smoothly connected) or discrete (non-continuous, incremental). Later, we will decide which view is the correct one.

To be sure, this question about time is not a new question such that we may reasonably assume that this question has been around so long that its origins are lost in the distant past. In fact, over the last two thousand years or so, we find that this question has been asked many times by individuals both famous and not so famous. For example, a classical and well-known example dating from the 5th century is a quote by St. Augustine (Christian-Roman Theologian, 354 – 430), which goes something like this; '*What then is time? If no one asks me, I know: if I wish to explain it to one that asketh, I know not.*'[8] A lot of us can relate to that, not only regarding time, but also to other common things of everyday experience.

The ancient Greeks, ever practical, are said to have seen time as a man riding backwards on a horse; he knows the past, but not the future.

Of course, not just time has been an ongoing topic of discussion throughout recorded history. Time has a companion called space that, over the centuries, has also been the topic of much discussion and debate. Anyway, of the two, time is more common in ordinary daily human activities. We use the word time in many ways, but seldom, if ever; do we think about what the word really means? Standard dictionary definitions are only somewhat helpful since they usually define the word time by means of everyday usage and often in a circular fashion. Dictionaries tend to use phrases such as *on time*, *good (or bad) timing*, and

---

[8] St. Augustine of Hippo (354-430), *Confessions of St Augustine Book XI* (New York: Modern Library, 1999), 123.

other phrases that are part of our normal daily conversations.[9]

Common Time

"So long as we do not go outside the domain of consciousness, the notion of time is relatively clear. Not only do we distinguish without difficulty present sensation from the remembrance of past sensations or the anticipation of future sensations, but we know perfectly well what we mean when we say that of two conscious phenomena which we remember, one was anterior to the other; or that, of two foreseen conscious phenomena, one will be anterior to the other." [10]

As stated in this quote by Henri Poincaré (French mathematician and philosopher, 1854–1912), if we stay within *the domain of consciousness* there is nothing uncomfortable about the notion of time. This is the common time of everyday experience, as influenced by past memories, current events, and the anticipated future.

Common time (aka naive time) is the time of our perceptions and it is a learned intuition of psychology and not the *a priori* intuition of

---

[9] For example, in a particular dictionary I found 64 usage definitions for time, and 24 for space.

[10] Henri Poincaré, (1913), "The Measure of Time," The Foundations of Science (New York: Science Press, 1913) pp. 222-234, <http://www.archive.org/details/foundationsscie01poingoog>

philosophy. That is, common time is a learned concept and not a natural intuition.

Common time is the time we experience in our everyday lives where, although we depend on clocks keeping a steady pace, our perception of this pace can vary greatly. For example, it is trivially obvious that we are all familiar with the difference between an hour spent under stress and an hour spent pleasantly entertained.

Natural Time

"Human perception is directed to the world; animal perception is directed to an environment."[11]

"Non-rational creatures do not look before or after but live in the animal eternity of a perpetual present; instinct is their animal grace and constant inspiration; and they are never tempted to live otherwise than in accord with their own animal dharma, or immanent law."[12]

Imagine a world without clocks or calendars. No hours, days of the week, months, or any of the other many constructions and artifacts we use for keeping track of what we call time. Imagine a world where the universe was much simpler than today – just the earth with its daily passage of the sun, the longer cycles of the moon and the nightly

---

[11] Maurice Merleau-Ponty, *The Primacy of Perception* (Evanston, IL: Northwestern University Press, 1964), 40.

[12] Aldus Huxley, *The Perennial Philosophy* (New York-London: Harper Modern Classics, (1945) 2009), 140.

spectacle of innumerable stars that appear to rotate about a fixed point.

This is the world run by natural cycles and is the world of natural time. This *time* is unfettered by human influences. It is the world of natural creatures today and that of humans in the dimness of prehistory, before becoming obsessed with time. Consider the following.

In the dead of winter in Antarctica hundreds of male Emperor Penguins will spend about 64 days huddled together while enduring fierce gales and frigid temperatures. Each male penguin does this for the sole purpose of incubating a single egg. During this period of sunless days, these male penguins have no food and they will lose almost half their body weight!

How are the penguins able to do this? Put simply, they can do this because their sense of time is very different from our sense of time. Driven by natural cycles, which give them independence from the detail schedules followed by most humans, they have a sense of natural time that is free from the anxieties of an anticipated future and regrets of the past. Put another way, as the second of the above quote states, animals live in the present, while, on the other hand, humans live in the *past* and *future* which are separated by a fleeting duration that we sometimes call the *present*.

For humans, such a task would be impossible. Even if we could endure the frigid weather and the lack of nourishment, our sense of time would render us incapable. We live by the clock, in a state of almost constant anticipation, and would soon find the sense of slowly passing time unbearable. On the other hand, instinct and the force of necessity are the drivers of the natural world, not clocks. In the natural world, there is great patience.

The natural cycles of day and night along with the lunar cycles and annular seasons have considerable influence on the natural world. However, this is not a human schedule driven by a clock, but is the union of perceived natural cycles with the creatures' physiological and psychological needs. In other words, natural time is *event-driven*, while our world is *schedule-driven*. We humans divide the day into hours, minutes, and seconds, then impose a schedule on ourselves that has little or no regard for natural cycles.

All life forms have an innate capability that is a guide for appropriate action. That is, all life forms possess, at a minimum, capabilities necessary for survival. Survival dictates that all life forms must interact and alter their immediate environment in pursuit of sustenance, reproduction, and ultimate survival. This is driven by the force of necessity such that pursuit of survival encompasses all living organisms from the great creatures of the ocean depths, to the smallest microbes. All life goes on within an environment. That may seem obvious but within our modern cities, this is often ignored. For example, the natural environment and the true sources of sustenance are filtered out by the system of supply that provides us with products that bear little, if any, resemblance to their sources.

The interrelationships between the environment and humans is, obviously, quite different from that of the other creatures. Although, all living creatures rely on a compatible environment for survival, it is only the human creature that imposes their will and makes unnatural demands on the environment. All other creatures work in and around their environment, while humans challenge and attempt to modify the environment with an attitude that seems to imply that *they* are the masters and not the natural world. This attitude is only adjusted or temporarily suspended by natural disasters and other natural occurrences over which we have no control.

Common Space

"Space is the boundless, three-dimensional extent in which objects and events occur and have relative position and direction. ... The concept of space is considered to be of fundamental importance to an understanding of the physical universe. However, disagreement continues between philosophers over whether it is itself an entity, a relationship between entities, or part of a conceptual framework."[13]

"Space: The physical universe beyond the earth's atmosphere. The near-vacuum extending between the

---

[13] Wikipedia definition (2018).

planets and stars, containing small amounts of gas and dust." [14]

"With Kant, space is given as a ready-made form of our perceptive faculty, a veritable 'deus ex machina'[15], of which we see neither how it arises, nor why it is what it is rather than anything else…"[16]

Most people who have every thought about it probably think of space in a manner similar to that specified in the first two of the above quotes. That is, as being a three-dimensional extension in which things exist and events occur. That is, space considered as something like a container or box. I will call this *common space* and as with common time, we feel that we have an intuitive understanding of common space.

However, common space is different from common (linear) time; such that our notions of common space is a cooperative composite of individual sensory spaces. These individual spaces include visual, tactile, and motile spaces, each

---

[14] Online Oxford English Dictionary definition (2018).

[15] *Deus ex machina* is a theatrical device frequently used by ancient Greek dramatists. It is a technique that provides for solutions to seemingly insolvable situations by the introduction of some arbitrary intervention by a god-like character that appears out of nowhere. This technique is often found in science fiction and fantasy writing. For example, J. K. Rowling's *Harry Potter* novels frequently use this dramatic device. It translates as *god from a machine*, which refers to the crane or other mechanical device used by the Greeks to make the *god-like* character suddenly appear.

[16] Henri Bergson, *Creative Evolution* (London: MacMillan and Co, 1911), 216.

with a different interpretation of space but all three are integrated into our overall notions of common space.

Visual space is the space of the images produced by our sense of sight and is a two-dimensional projection on the retinae of our eyes. These images, having only height and width, lack a third dimension. The additional dimension of depth is a mental construction using the convergence of sensations in our two eyes.[17] That is, our perception of the third dimension (depth perception) is a mentally constructed interpretation.

Tactile Space is the space of our sense of touch. It assists in producing our notions about the reality of extended matter[18] and contributes to our impression of the world being three-dimensional. Our tactile sense provides the sense data for building tactile space through a combination of the texture of objects and the feel of extension of these objects. For example, writing this sentence involves contact with the pen I am holding such that I am aware of the pen's texture and extension. Tactile space also plays an important role in

---

[17] Herein, I define sensations as the initial detection of external stimulus by a sense organ.

[18] Matter, as defined herein, includes not only extended persistent objects, but also includes energy and matter in all their known forms as well as any hidden structure and composition not yet discovered. Perhaps it should just be called the *stuff of the universe* or simply *world-stuff*.

helping us to establish the reality of objects. To touch an object is to believe it real.

Motile or motor space is the third space that we learn as infants. We slowly learn to control the movements of our body parts, starting with jerky and uncertain movements we manage to progress through a process of trial and error to a level of proficiency that enables controlled and purposeful movement. This space not only helps us to move about and manipulate objects, but also helps us build a sense of direction and awareness of our immediate spatial environment.

Common space is an integration of these subtypes, which enables us to function in what is primarily a two-dimensional environment. Sure, we sometimes go up or down, but for the most part, we move in two dimensions with the third dimension being generally limited to providing perspectives of size and distance, which helps us not walk into walls or other things.

Not only are the senses our window into the external world, but they work together to confirm our belief in veridical reality. For example, when we observe a simple cylinder, such as a can of beans, from one perspective we observe a rectangle, while from another we see a circle. Through rotation we observe the transition from circle to rectangle, and back again. The confirmation of the cylinder comes from our integration of the tactile sense, such as feeling the roundness and through visual sensations, assisted by the motile sense, which allows us to manipulate objects to gain a more complete image.

As we progress from infancy to maturity, our perceptual conceptions of space evolve and expand into our notion of common space. In other words, we build our notions of common space through the trial and error of perceptual experience.

Preliminary Topics

These next few topics are here to provide some information that should prove useful both individually and as background information for the rest of the book.

The titles of these topics may seem technical, but they are presented in non-technical fashion to more easily facilitate conceptual understanding of these topics for the non-specialist and how they relate to the later topics presented herein.

Concept Space

"Nothing is presupposed save the existence of things in their inexhaustible multiplicity, and the power of the mind to select from this wealth of particular existences those features that are common to several of them."[19]

"…concepts do not gain their truth by being copies of realities presented in themselves, but by expressing ideal orders by which the connection of experiences is established and guaranteed."[20]

---

[19] Ernst Cassirer (German philosopher, 1874 – 1945), *Substance and Function*, (New York: Dover Publications, 1953), 4.

[20] Ibid, 319.

"Concepts are there, but not in any definite place. They ...form the 'conceptual space'."[21]

Concept space is the mental space of concepts constructed through perceptualizing our encounters with objects and events. Although the process of forming concepts and use of concepts goes largely unnoticed, this concept space is indispensable for normal functioning.

Most dictionary definitions define concepts as ideas or images that are abstractions, mental impressions, or notions. Such definitions are correct, as far as they go, but they are also incomplete and generally fail to do justice to the importance of concepts. More than simply mental impressions or representations, concepts are abstract generalizations, and associative correlations that provide an essential capability for dealing with the continuous flow of complex sensations and perceptual multiplicity.

Concepts are both efficacious and essential elements of our capacity to handle the constant bombardment of massive amounts of sense data. Through a process of abstraction, summarization, and classification, concepts facilitate our ability to assimilation copious amounts of data as generalized patterns, templates, and abbreviations.

---

[21] Kurt Gödel [Austrian-American logician, mathematician, and philosopher, 1906 – 1978] quoted in: Palle Yourgau: *A World Without Time*, (Cambridge, MA: Basic Books, 2005), 175.

Put simply, concepts are indispensable short cuts for rapid recall, decision-making, and just plain normal functioning. Only through building and using concepts are we able to function as we do. Concepts play a necessary role in essentially all aspects of our lives, but perhaps the primary use of concepts is to facilitate accommodation, assimilation, and classification of phenomenal experience.

A concept is an association with the vanished content of past objects and events, which lacking the exactness of these past objects and events, is often only a blurred remembrance. A concept is not that which is perceived (the percept), and it is not possible, to effectively visualize a concept. That is, whether a concept has only one referent or a multitude of referents, the concept itself is not one of the referents.[22] That is, the concept itself is not a member of the concept's set of referents and cannot be instantiated or exemplification. Attempting to visualize a concept will only result in the image of a known particular or referent.

Concepts are also material for reflection and contemplation. New concepts are synthesized through contemplative reflection and association of *similar-differences* and *different-similarities*. They mediate between thought and language such that they assist in the creation of new levels of

---

[22] A referent as defined herein can be a physical object, an event, a fictitious thing, or a thought (such as an emotion).

knowledge. Through concepts, there emerges levels of meaning external to and independent of the sources of this extracted knowledge. That is, concepts, being removed from place and time, are available for free associations such that they assist in the building of new inferences and new content. Concepts create categories *of* and *for* understanding.

Concepts can be divided into the two basic categories of *thing-concepts* and *event-concepts*. Thing-concepts are the first to be learned and are somewhat self-explanatory, while event-concepts, as the name implies, are associated with actions – especially repetitive actions – and are basic to both humans and the higher animals. Examples of this type include such everyday activities as writing and driving an auto. However, they also have a higher level that include processes such as interactions within family and non-familial social interactions. Put another way, event-concepts include most of what we do every day and are learned through our interactions with others. Together, thing-concepts and event-concepts being similar in structure may be considered in combination as thing-event concepts

Types of Concepts
Although, there are many ways to classify concepts, the two categories just mentioned can be sub-divided into three sub-categories or types that are sufficient to include most, if not all. These three sub-categories will be called: *attributive, generic*, and *individual*.

Attributive: These concepts are probably the type most recognized and represent the simplest type of concept. As the name implies, this type of concept represents the phenomenal attributes or properties of objects (the referents) that are used to define and identify the object. Examples of attributive concepts are many and include such notions as *red, furry, biped,* and so on.

Open any dictionary and select the definition of any object. It is certain that the definition chosen will refer to attributes of the object and that the definition assumes a basic understanding of these attributes. Attributive concepts, of course, are not restricted to properties of objects, but also include properties of events, such as *sunny*, *cold*, *outdoor*, *birthday party*, and so on.

It might be argued that attributive concepts are not true concepts, but simply attributes or characteristics. However, consider the concept of a particular color, since the concept itself cannot be visualized, our experience of a particular color is never the conceptual color, but rather an example of the conceptualized color. For example, when we consider the concept *blue*, we visualize a particular shade of blue, which is an exemplification of the *blue* concept. That is, the *blue* concept is the standard by which we recognize and associate all shades of blue.

However, through adjectival modification, we extend concepts by concatenating them with other concepts. For example, the perceived blue, which is never the conceptual blue, might be *light-blue, red-blue* or any other shade of blue. Put somewhat differently, when we say *This is blue.* we have a relationship between a real object as perceived and the *blue* concept.

Generic: These concepts represent the aggregated properties of objects. That is, these concepts are composite concepts and always lack the simplicity of attributive concepts. These concepts always have multiple referents. For example, the generic concept of *tree* is the aggregate of the attributive and generic concepts common to all trees and supervenes upon the attributive and generic concepts of specific types of trees. As the name implies, generic concepts are generalizations and it is through these generalizations that we build our conceptual hierarchies and notions of universals.[23]

Generic concepts form categorical levels such that hierarchical relations exist between these levels. Concepts of a lower (more specific) levels are inclusive within the higher (more general) levels, while the lower levels are exclusive of all other concepts included within the same level. For example, the *pine* concept is included in the more inclusive generic *tree* concept and is exclusive of all other types of trees – such as the *oak* concept, while the concept of *tree* is included in the more inclusive generic concept of *plant*, such

---

[23] It should be noted that all events are conceptually generic. Every event is a complex set of objects and action, including the observer and the observed.

that within the hierarchy, all pines are trees and all trees are plants. On the other hand, the tree concept excludes all non-tree plants.

The hierarchy of generic concepts relies on either adding or deleting various attributes (or properties). Adding attributes leads to greater specificity, while deleting attributes leads to greater generality. As mentioned above, at one extreme is the specificity of but a single exemplification, and at the other end of increased generality is a categorical apex such as the concept of *plant*.

This is not to say that all reach the same apex, but rather due to category scope and the complexity of conceptual thought each category eventually reaches an apex or terminating node. For example, both *plant* – includes all plants - and *tree* – includes all trees - concepts are apexes within their respective hierarchies and if specific attributes are removed from the *tree* concept, it would eventually generalize as the *plant* concept.

Individual: As the name implies, this type of concept represents only a single thing or referent. It might seem strange to consider an individual object as a concept; however, when thought about in a certain way, such a notion, or should I say concept, is easy to accept. For example, each human individual exemplifies the attributive and generic concepts associated with all humans, plus the attributes that uniquely individualize this particular person. For example, an individual named Napoleon, which as the *Napoleon* concept is dependent on a multitude of concepts such as, *human*, *biped*, *Corsican*, *French Emperor*, *married to Josephine*, and so on.

Every person we know is an *individual* concept. Furthermore, each person knowing the same person has their own unique concept of this person that is customized based on relationship, past interactions, personality (of both persons), and so on. Individual concepts have no restrictions on referents except, as the name implies, individual concepts require that the individual thing (person, animal, object, event, or thought) have a uniquely discernable identity.

Non-real Concepts

What has been discussed above, primarily concerns *thing-event* concepts, which reference *real* objects and events. However, in this section, three types of *non-real* concepts are considered: *intimate*, *fictional*, and *relational*. Although, most of what has been discussed also applies to non-real concepts, these concepts, which might be considered exceptional cases of generic concepts, are different enough to warrant a separate discussion. Since all concepts are abstractions existing in concept space, it might seem strange to refer to non-real concepts. However, non-real concepts are those that have no direct connections with concrete objects or actual events. That is, the referents are unreal.

For example, the referents of *intimate* concepts exist only as mental constructions, however, the construction of these concepts relies, at least partially, on personal experience. For example, the emotion of sadness or the sensations of pain are learned through experiencing events that evoked an emotional and/or psychophysical response and it is the repetition of these experiences that leads to the construction of *non-real intimate* concepts such as *sadness* and *pain*.

Although, often initially motivated by actual events, as these concepts mature, they have a tendency, except in the cases of extreme trauma, to become disassociated from these events and exist as purely mental notions. This type of concept includes not only emotional states but include any thoughts that do not readily lend themselves to visualization or verbalization. For example, the intimate concepts associated with religious or political symbols.

On the other hand, *fictional* concepts are non-real concepts that relate to imaginary objects or events. For example, the concepts of *unicorn* and *Captain Ahab* are fictional concepts. Although fictional, these concepts rely on real *thing-event* concepts. For example, the *unicorn* concept relies on the concepts of *horse*, *horn*, and so on, while the *Captain Ahab* concept relies on the concepts of *human*, *ship's captain*, and so on.

The third type of non-real concepts to be considered are the *relational* concepts. These concepts are about comparisons or associations between two or more things and are inclusive within the broader concept of *order*. Even though there are many types of order, commonly, we tend to think that order has something to do with the systematic arrangement of physical objects or a particular way of doing things. Our conceptual understanding of order is not intuitive but is one of the first things we learn [*Order*, page 103].

Starting in early childhood, parents and others teach us the concept of ordering things. First, just by lining things up and later as we develop language skills, our sense of ordering becomes more complex and concepts like *alphabetical order* and *numeric order* become familiar. The referents of relational concepts can be either objects, events, or pure thoughts (such as emotions), but the relational concepts themselves are context free. For example, the notions of *greater than*, *before-after*, *friend of*, *in-between*, and *happier than* are conceptual relations, which are contextless and available for free associations.[24]

Non-verbal Concepts

Apparently, there are those who consider linguistic capability essential for the possession of concepts and apparently believe that verbal (word-like) concepts are the only kind that can exist.

---

[24] Mathematics is essentially a system of relational concepts where the syntax is defined by axioms, definitions, and other rules of the game. For example, consider this simple equation: $1 + 1 = 2$. This equation incorporates a number of non-real (abstract) concepts such as '1', '2', as well as the concepts of addition (+) and equality or equivalence (=).

This ignores the fact that all pre-linguistic children (and non-human higher creatures) have what might be called an inchoate or rudimentary ability to construct concepts. That is, pre-linguistic humans and creatures capable of true learning have an innate ability to abstractly process perceptions by selectively extracting similarities (and differences) without overt conscious effort. And, in a manner similar to the acquisition of motor skills, this innate ability must be exercised to improve functionally such that as the individual matures, this ability also matures, and concepts become more complex.

Also, we should consider whether there is a difference between human concepts and what some call *animal representations*? As far as we know or are able to determine, the difference between humans and the higher creatures is that only humans take this process of abstraction further with the contemplation of concept themselves. To say that animals, being non-linguistic, only have mere representations is just another example of anthropic thinking and the making of unwarranted assumptions about human uniqueness.

Associating non-linguistic capability with non-conceptual representational thought (mental states?) and linguistic capability with conceptual states seems suspect since there appears to be no physical or cognitive reason for this belief. That is, assuming non-linguistic humans and animals are only capable of non-conceptual states has no physical or mental basis and arbitrarily precludes any pictorial or other non-verbal concepts such as those associated with our olfactory sense. For example, sweet and sour are learned concepts well before acquiring verbal representations of these sensations.

By necessity, these non-verbal concepts are abstractions or pictorial representations of audile, tactile, visual, olfactory, and motile sensations; however, regardless of the form of these concepts, they apparently serve their purpose. For example, both pre-linguistic humans and higher creatures are known to react to signals in a manner that indicates learned generalized notions and these generalizations indicate conceptual thought. For example, all animals immediately recognize others of their own kind, which implies that they have something like a concept of their species.

We learn about objects and events before we learn language and it is from these early experiences that we build non-verbal concepts – images such as a toy (nameless but recognizable) and non-image concepts such as pain – that are associative mental constructions of these objects and events, which, later, after the acquisition of language they become associated with words. But even then, we still maintain non-verbal concepts where either a corresponding word does not exist or is unknown to the person.

Moreover, denying non-verbal concepts ignores the fact that even humans with language skills do not actually forgo all non-verbal thought.

Thinking in non-linguistic images is not only common, but also essential because of the limitations of language when handling complexity or ineffability that invokes strong non-verbal emotions such as those associated with religion and other highly emotional human issues.

Summary
When we put it all together, we see that generic and individual concepts are truly multi-dimensional; that form an n-dimensional space where n can be a very large, but finite number that increases with the complexity of the concepts and completeness of our knowledge. When we think of a complex thing, we do so via a set of concepts, which constitutes our knowledge of this thing both generically and specifically.

Put another way, concepts are much like building blocks for associations and classifications that enable us to breakdown the unfamiliar by comparison with the familiar – or put still another way, concepts provide the necessary components that enable us to quickly discern perceptually encounter similar-differences and different-similarities. It is how we are able to quickly recognize both normal perceptions and the unexpected perceptions such as an unexpected hazard.

For example, as mentioned previously, we know other humans by their generic concepts that are common to all humans and by the individual attributes possessed by only a single instantiation of the *human* concept. However, due to the limitations of human capabilities, conceptual associations of complex entities such as humans are always incomplete.

Without concepts, every object perceived. event encountered, and even thoughts would have to be individually analyzed and remembered. Learning as we know it would not be possible and our ability to perform most, if not all, of the tasks that we perform daily would be seriously impaired, if not impossible. Without doubt, it is difficult to overstate the importance and indispensability of concepts.

Concepts allow us to extrapolate beyond the given world of perception. They provide us with the indispensable tools and capabilities necessary to think at levels of complexity that would otherwise not be possible. Just as mathematical equations are capable of *compressing* complexity into a few symbols (that are themselves concepts), concepts compress information and knowledge into manageable pieces that might be called intellectual abbreviations. Essentially everything we experience and contemplate involves concepts. These very words are concepts, the act of writing is a concept, and the thoughts expressed in speech and writing are concepts.

Perception and Reality

Although at first glance, perception and reality may not seem related to space and time discussions, however since what we perceive effects our reality and, in a circular manner, our version of reality influences how we perceive the external world. That is, it is through perception and our version of reality that we build up our notions of space and time.

Common Perception

Although a constant companion while awake, the process of perception is mostly ignored. Even now while reading these words, the process of perception, although probably not noticed, is at work. Only in dreamless sleep or a state of anesthetized unconsciousness do we escape this constant companion. Not only is perception persistent, it is our window into the external world.

Common perception, which might be called *common sense* perception, is something we use in a manner like the way we use technology. Although we use technology on an everyday basis, for the most part, we only have *operational* knowledge of this technology. Operational knowledge is like following a recipe in that recipes provides necessary steps for construction but no functional explanations such as why this and why that.

That is, we know how to use technology, but have little or no *functional* knowledge of technology. For example, the use of personal communication devices (aka cell phones) is common to the point of being universal, but very few users have even the slightest idea of how cellular communication functions. The same is true of perception, our perception is always active and in use, but our awareness of the process of perception is for the most part hidden and unknown.

During normal daily activities, we do not consider or question the details of perception, we simply accept what we perceive and let it go at that. Of course, this is very practical, since if we stopped to examine the details of the flood of received sensations, we would quickly become overloaded and find ourselves incapable of normal functioning. Normal activity requires that we not spend too much time analyzing our perceptions.

Perception as commonly experienced assumes the reality of perceived objects. We intuitively feel that these objects have an immanent existence of their own just as we feel the reality of our own existence. What we usually do not consider is that it is through perception that we are able to build up our notions of reality and confidence in the reality of the world.

Side Comment on Sensation versus Perception:[25]
Here is an example to help illustrate the difference between sensation and perception. When a young child or infant first encounters lemonade, they have a sensation of the taste of the lemonade as a simple undifferentiated sensation. With experience and more maturity, the lemonade becomes a compound sensation - a complete perception. That is, as children gain experience, they begin to differentiate the lemonade into components such as sour, sweet, and cool. The initial sensations of lemonade transform into the more complex and complete perception of lemonade.

Common Reality

Herein, the definition of common reality is the reality of common or everyday perception. Although somewhat circular and not a particularly useful definition, it is more or less accurate. Put another way, reality is centered on ourselves, that is, our complete self, both physical and mental (the psychophysical self). Everything we experience is a projection onto and through ourselves; and it is through reception of these projections that we build our reality and total view of the external world. For example, while writing this sentence, there is the sense of contact with the computer keyboard, which together

---

[25] I use side comments when either when there is only an indirectly connection with the current topic, a more technical note, or just when I felt like it.

with the computer and the immediate surrounding, is considered real.

These things are *real* because they are persistent and familiar. That is, confidence in the reality of *ourselves* and the *not-ourselves* is built up through perceptual persistency of objects and the familiarity of recurring events. It is our confidence in this common reality that carries forward into the future with expectations of consistency that are essential for our sense of well-being and normalcy.

Commonly, we usually do not differentiate between the real and reality since in common reality they essentially coincide, and it is only when we dig deeper into these concepts of real and reality, do we find that they are indeed very different notions.

Reconsidering Perception and Reality

After briefly discussing some aspects of common perception and common reality, it is time to reconsider these notions in a manner more analytic and take a deeper look beyond the common notions of perception and reality.

As this process proceeds, we shall find that both are more complex and interconnected than normally considered. In fact, although both are discussed separately, I hope to show that perception and reality are concomitantly connected and that any discussion of one necessarily entails the other - either explicitly or implicitly.

Perception Reconsidered

"But even the places in which I find myself are never completely given to me; the things I see are things for me only under the condition that they always recede beyond their immediately given aspects. Thus, there is a paradox of immanence and transcendence in perception. Immanence, because the perceived object cannot be foreign to him who perceives; transcendence, because it always contains something more than what is actually given. And these two elements of perception are not, properly speaking, contradictory."[26]

"...the senses are the organs through which the live creature participates directly in the on-goings of the world about... In this participation, the varied wonder and splendor of this world are made actual..."[27]

"It is in reference to our own body that we locate exterior objects, and the only special relations of these objects that we can picture to ourselves are their relations with our body. It is our body that serves us, so to speak, as a system of axes of co-ordinates."[28]

The authors of the above quotes clearly recognized that perception is not merely our *window on the world* but is the fundamental source of knowledge and human experience, and that this *window* is, through our sense organs, centered around our bodies. Therefore, expanding our

---

[26] Merleau-Ponty 1964, 16.

[27] John Dewey (1934), *Art as Experience*, (NY, Penguin Books, 2005), 22.

[28] Poincaré 1918, 100.

understanding of perception beyond the elementary (or common) level provides the opportunity for both increased self-understanding and enhancing our worldview.

As stated above, perception is immanent because it is presented to us in a very personal way and that which we perceive is believed to have a concrete reality (concrescence). On the other hand, perception is transcendental because beyond the reality of that perceived, the thing-in-itself remains unknown and, perhaps, unknowable. That is, our perception is always incomplete; we only have phenomenological knowledge of objects and events. Even if we consider this knowledge empirical or scientific, we cannot pass beyond the phenomena except by speculative inference and extrapolation.

The notion that through perception we live reality as experienced is another way of stating the primacy of perception. Perception involves a transcendence of the physical percepts into the mental realm of ideas, concepts, and thoughts. When an object or event is committed to our perceptual storehouse, the physical object is not directly changed, but its phenomenological image is integrated into the Mind where it is wedged in with a myriad of other mental occupants.[29]

---

[29] Herein *Mind* (with upper case M) is a reference to all psycho-physical components, including the subconscious, and the non-

Beginning with our earliest sensations of the *not-ourselves* external world, we begin the process of receiving and processing signals from outside of ourselves. Initially these signals or sensations seem incomprehensibly random and uncorrelated, however, over time, we begin to make sense of these signals through an ability to recognize repetitious patterns and retain images of these patterns in the form of concepts.

Although perception seems like an objective process and can be broken down into objective steps, after the innocence of infancy there are always subjective influences that modify *pure* perception.[30] The sources of these subjective influences are both social and personal. Society, both globally and locally, influences our perception through convention, peer influence, social expectations, and so on, while at the internal or personal level, our perceptions are influenced by how we interpret our environment, our close associations, and by our unique personal experiences.

There can be no doubt that experience is prior to and necessary for our perceptual notions of

---

physical aspects of consciousness, while mind (with lower case m) refers primarily to common mental functions such as awareness.

[30] Herein, the *objective* is considered a reference to something independent of a perceiving mind, while the *subjective* is somewhat the opposite and refers to some notion held in the thinking mind. Put another way, objective is what we believe as having independent reality and subjective is a personalized interpretation.

external reality; and prior experience is gained through sensory experiences. It is only through the acceptance of perceived objects and events as being something we call real that we know and acknowledge external reality.

Our experience of the world is mediated through reconciliation of active sensations with prior perceptual experience. As the experiences of the past are brought forward for correlation with the current experience, the past is projected into the current experience. The immediate sense data are correlated with memory of similar content. However, this correlation is never exact such that the current observation of a familiar object is always different from the previous view. Something has changed, be it ever so slight.

To maintain comfort and normalcy, we tend to seek our reality by the least effort path and avoid the agitation of a difficult path. This applies both to our physical as well as to our mental comfort, since without both there is neither.

Contrary to other forms of life, where the focus is on sustenance, reproduction, and survival in an environment, human activities include contemplating the utility of things and events, as well as seeking meanings for these things and events. That is, the human mind reflects upon itself (self-reflexive) and ponders the world in a general way. As far as we know, or are capable of knowing, only humans pursue endeavors that seek the function and meaning (or purpose) behind the phenomena.

Although there is, in general, consensus concerning our common perception of objects and events, there is always some differentiation as well. Each of us has a different understanding and interpretation of perceived objects and events that reflect our personal preferences, experiences, and general inclinations. For example, when I first see a novel table, without conscious effort, I immediately correlate this table with personal encounters of other tables. These prior encounters with other tables have through a process of synthesis and abstraction developed into a generic image of table, which might well be called a conceptual notion of *table-hood* and through this process each of us has a different notion of this table-hood. For example, a furniture artisan will *see* a table in a more distinctive way than most of us.

That is, perception has common elements based on consensual intersubjective agreements by society at large, as well as personal elements unique to the individual. Put another way, perception has an interpretive aspect based on interest, education, inclination, and so on. Even science is only an extension of what we do in our everyday lives. What makes science different is the use of formal procedures and rules to extend this synthesis and make it more precise. In everyday situations, affirmation of the table as a table is immediate and easily understood, while for the objectives of scientific study, determination, and affirmation, are more drawn out and more metrically exact.

In summary, perception is not just a common flow of objects and events. The perception of an object is individualized, with each person having an individual version of the object perceived. To say that the astronomer sees the moon more accurately that a Bushman of the Kalahari is not a statement of value, but a statement about a technological viewpoint versus what might be called a natural or spiritual viewpoint.

Reality Reconsidered

"Physical reality, then, is the realm of actual and possible sense perceptions. The concretely physical is just that which yields and sustains perceptual experiences."[31]

"The belief in physical reality is really, in a final analysis, belief in a public realm of experience, accessible to other percipients of like nature with one's self. This belief, therefore, rests on the recognition of a social realm of beings with the same perceptual and rational powers."[32]

---

[31] J. A Leighton, "Perception and Physical Reality." The Philosophical Review, Vol. 19, No. 1 (Jan. 1910), pp. 1-21, Published by: Duke University Press on behalf of Philosophical Review, URL: http://www.jstor.org/stable/2177636, 10.

[32] Ibid, 16.

"The aim of the whole process seems to be to reach as complete and united a conception of reality as possible, a conception wherein the greatest fullness of data shall be combined with the greatest simplicity of conception. The effort of consciousness seems to be to combine the greatest richness of content with the greatest definiteness of organization."[33]

As the above quotes state, our reality has both personal and collective aspects. That is, reality based on personal experience of the world as presented and interpreted by our senses and on community agreement. Since perception is the primary source of our reality, understanding perception is necessary if we are to understand our reality. Experience of the external world through accumulation, accommodation, and assimilation of sensations forms the basis for reality. The experiences of external sensations are expanded through the mental processes of extrapolation, induction, deduction, and so on. Although, not completely reliable, these processes are the only way we gain knowledge of the external world.

Besides the word reality, we also use the word *actual* to designate what we feel is real. However, the word actual has a somewhat different usage, as it is often used to emphasis current existence, a recent occurrence, or an immediate presence. For example, statements such as *I was actually*

---

[33] Josiah Royce, The Religious Aspects of Philosophy (Boston and New York: Houghton Mifflin Company, 1885), 357.

there. indicate an emphasis on being personally present.

Consider also the words reality and real. Reality is the world of phenomena and our perception, real is the world behind the phenomena and perception. Reality is the thing as perceived, real is the unknown and unknowable thing-in-itself.

The greatest reality is the reality of our own existence. It is the *I*, which dominates us all and is the center of our world. Although this sounds egotistical, it is not meant in that vein. This is just the reality of existence and how could it be otherwise? The humblest and most selfless person is still the center of their own world (existence).

This is *self-centric*, as opposed to a *self-centered* and is inescapable. With either greater or lesser degrees of centricity, the personal self is always present in evaluation of sensation and determination of action. Traditionally, self-denial was (and is) the path of all saints of all religions, however, today, such self-denial seems impossible and probably is given the *"it's all about me"* attitude that seems to be pervasive with our society Put simply, with the rise of the *selfie generation*, it seems that self-centered has become dominant.

Types of Reality

Within the technical literature, of the type that discusses such things, we encounter a plethora of realities and realisms. Each was conceived to meet a perceived need and each version of reality has subscribers in agreement with that realism.

Having said that, a reader believing in only one kind of reality might be inclined to object. Certainly, on the surface it seems that only one reality, often called common (or naive) reality, exists. However, when we dig deeper into what we normally consider *reality*, we find that it does have many variations.

In no particular order, here is a short, but representative sampling of the various realism that might be encountered, Conceptual Realism, Scientific Realism, Conventional Realism, Platonic Realism, Empirical Realism, Common Realism, and Operational Realism. This list could go on, but as an example it is enough to get across the idea that there is no fixed – one size fits all – reality.

The events in our lives manifest themselves in two distinct, but connected, ways. The natural world of phenomena, for the most part, is objectively presented; however, our interpretation and contemplation of these events is subjectively influenced and individually unique. Neither the objective nor the subjective can be avoided and together they determine our version of reality. When the objective experience - the phenomenal events common to all - and the subjective interpretation meet certain cultural and social standards, then the observer is said to be normal within the constraints and conventions common to the observer's culture.

Stated simply, reality has a certain amount of variability, which depends on such things as individual capabilities, knowledge, social environment, inclination, and so on. To illustrate this point, here is an example of two very different realities.

In the Kalahari Desert of southwest Africa, reside the native Bushmen (San People) who, from the point of view of environment, technology, and formal education, live very primitive lives. Indeed, until recently, these people had lives not far removed from what we commonly call stone-age humanity. At the other extreme are the typical young urban dwellers, perhaps living in Los Angeles or London, who are constantly connected to the latest technical gadgets, especially those starting with '*i*'.

These various technical accoutrements tend to be constant companions of urban living in general and youthful urban living in particular. On the other hand, the *advanced* technology of the Bushmen is fire generated by wood friction and a bow with arrows. These individuals live in habitats about as far apart, regarding both social environment and technology, as is possible these days.

However, these desert dwellers, although technologically primitive, have an intimate alignment with the natural world, which includes not only physical objects, but also includes a strong spiritual sensibility and an awareness of the integrated wholeness of existence. In contrast, the urban dweller, technologically over equipped and spiritually deprived, is mainly occupied with technology and a certain social behavior mediated by current fashion.

The San People are fully integrated into the natural world, while the urban dweller is fully integrated into the world of artifacts and superficiality.[34] It is not hard to imagine that each have a very different view of reality. What one is missing, the other has in abundance.

To be sure, we do not have to go to such extremes to find variations in reality; each of us has a personal view that is never exactly the same as that of another. For example, two persons viewing the same rainbow will probably have different interpretations of the event. A meteorologist, for example, will look beyond the phenomenon and see not only the rainbow, but will reflect on the physical causes of rainbows. On the other hand, an artist will most likely see the esthetics of the rainbow. Which interpretation is the *real* one? Of course, both are just as real for each person.

Reality is partially based on our assumption and belief in veridical perception. That is, perception based on or coinciding with what we judge to be real. As the observer's knowledge becomes more complete, in the sense of increased insight into the observed structure and properties of the

---

[34] Unfortunately, the government of Botswana has re-located many of the Bushmen, and their old way of living is quickly fading.

objects and events experienced, then the observer's reality is modified accordingly.

Of course, the object itself does not change; only the observer's image and knowledge of the object is changed. Put simply, seeking knowledge and increasing understanding modifies individual reality and enhances a person's worldview.

This is somewhat circular but what we perceive and understand depends on our worldview, while, in turn, our worldview depends on what we perceive and understand. Each of us has a private *universe* that is unique and overlays the physical universe as presented to our senses. This is true, in a proportional manner, of all creatures who by their capabilities and within the limitations of their senses and intelligence create their own private reality.

Knowledge enhances the capacity to *see* beyond the common perceptual perspective of things and perceive the wonders of nature with more depth and understanding. This acquisition of knowledge is different from the overuse and dependence on addictive technical accoutrements of modern living as mentioned above. Addictive dependence is not knowledge.

Perceptual consistency is fundamental to our notions of external reality. This consistency is achieved through the continuous correlation of active sensations with past sensations. This process is more than the mere collection of data but includes processing the data into information and information into knowledge. The reader will recall that although the stream of sensations seems continuous, our experiences of these sensations is episodic (such as the *Specious Present* – page 114) in the sense that we consciously perceive intermittently and with variable intensities.

The concrescence or the objective reality of an external object is most strongly determined through perceptual persistence and consistency. Multiple encounters with the same or similar objects reinforce the concept of the existence of a particular object and in this way, the external world as projected into us is integrated with our sense of self to form the totality of our reality.

All versions of reality, at least those considered rational, are built on a fundamental and unalterable basis of all realities. This basis of reality includes, for example, unalterable and unavoidable components of reality such as the irrepressible force of gravity, the limits to biological existence, biological necessities such as sustenance, and so on, and this basis necessarily forms the foundation of all variations of reality, even those that are not completely rational.

Given this unavoidable basis or foundation, one might question whether these *variations of reality* are nothing more than mental exercises. However, there can be no doubt that since we pattern our lives on our personal worldview, these variations must be considered truly a part of our personal reality.

On reaching adulthood, we are, for the most part, set in our ways and have established a reality for ourselves, which will influence all our activities and opinions. Although it is possible for a person to modify their worldview somewhat and suppress certain behaviors or features of their reality, a full escape is probably not possible.

At the inner most level, our personal reality is deeply ingrained such that changes are difficult or even impossible. We can change cosmetically or superficially, but just as the topiary designer may change its outer appearance, the bush is still the same bush, similarly, the person's basic nature remains.

Side Comment on Apple: When the Apple Company began using the lower case 'i' as part of their product naming conventions, they consciously knew that this naming convention would appeal to those who consider themselves as different – in the sense of feeling superior - from most of humanity.

In fact, even prior to this new naming convention, Apple advertisements from the very beginning – probably mirroring Steve Jobs extreme egocentricity – targeted those who think themselves special and a cut above the masses and looked down on us ordinary folk who use Windows. The timing for this reference to the 'i' was appropriate for the egocentric mindset of the *selfie generation* that is now highly prevalent in this country and around the world. If you walk by an Apple store in just about any mall you will find it full of persons just playing around with various accoutrements with apparent fascination, and if a new product or upgrade is announced a techno-frenzy is bound to follow.

Scientific Reality

"The objects of physics are thus, in their connection according to law, not so much 'signs of something objective' as rather objective signs, that satisfy certain conceptual conditions and demands."[35]

In general, prior to the development of quantum theory, objects of scientific study were limited, for the most part, to those existing in the macro-world. That is, classical science was mostly concerned with the world of everyday experience, which is, for the most part, directly observable.[36]

However, with the development of atomic and quantum physics, science moved into the micro-world where direct observation of the phenomena was no longer possible. The micro-world is the world of the unvisible and gives up its secrets only indirectly and reluctantly through subtle hints. Studying the micro-world is somewhat like determining the shape and size of a rock tossed into a pond by examining the ripples.

Additionally, atomic and sub-atomic physics has an interpretive problem in trying to translate the postulated phenomena of the unvisible micro-world into the language of the visible macro-world. The deepest and most advanced science

---

[35] Ernst Cassirer, 1953, 305.

[36] Classical studies of electricity and magnetism were of micro-phenomena but relied on visual macro-effects for explanation.

still requires discussion and explanation in common everyday language and this need has revealed certain linguistic inadequacies. That is, language is adept at describing the macro-world from a mechanistic, cause-effect, and deterministic point of view, but as the micro-world apparently has none of these attributes, attempts to describe the micro-world are difficult and never completely successful.

Scientifically probing the unvisible microscopic world reveals only reflections or shadows. These are the *objective signs* mentioned in the above quote, which are revealed to us only through amplification and extrapolations of signals from the unvisible world. This process requires a great many complex steps before reaching a level *observable* and interpretable. Furthermore, these signals from the unvisible micro-world lend themselves to more than one interpretation and, therefore, more than one theoretical model. Consequently, any preference for one model over another almost certainly has subjective elements based on personal inclination, motivation, and so on.

Because of limited linguistic capability and variable reliability of our sense organs, there will always be a separation between our scientific knowledge and the things-in-themselves. Nevertheless, science has been remarkably successful in its attempts to overcome our sensual limitations and progress beyond the phenomena in an unending quest for ultimate reality.

Scientific reality is actually a human invention that simplifies the laws of nature. We know somethings about these laws of nature such that we are able to build scientific theories to explain this partial knowledge of nature, but these simplified versions are incomplete, and we only have clues about the true laws of nature that lie hidden.

Summary: Perception and Reality

Since humans are part of the natural world, should not their constructions also be considered part of the natural world or do our constructions form a barrier separating us from the natural world? We will leave the answer to the first part of the question open and answer the second part in the affirmative. As previously stated, (page 39), technology in its many manifestations tends to close out the natural world both physically and spiritually.

It is only in the remote parts of the world, among the most technologically primitive cultures still existing do we find societies that are still connected to the wholeness of natural world. These societies are sometimes deemed superstitious in their relations with nature; however, these societies have spiritual connections with the natural world that technologically endowed societies have all but abandoned and for the most part forgotten.

We live in two worlds, the external material world of events, objects, and sensation; and the internal mental world of concepts, emotion, and thought. Through perception, the reality of the external world enters into ourselves and blends with our mental world. Through this process, and in life's experiences, perception and reality are inextricably joined.

Classical Science

Modern science, or science as we think of it today, began developing around 1600 through the efforts of a few enlightened individuals such as Galileo Galilei (Italian, 1564-1642) and Johannes Kepler (German, 1571-1630). Although both remained partially under the influence of alchemy and astrology, they were capable of going beyond these ancient occupations and introduce analytical elements into their investigations.

Their efforts were the beginnings of modern science. Kepler, for example, systematically analyzed a vast amount of astronomical data and develop kinematic theories about planetary motion that are still valid.[37] Following the examples

---

[37] Kepler used the measurements collected by Tycho Brahe (Danish astronomer, 1546 – 1601), who for about 30 years recorded a vast number of celestial observations.

of Galileo and Kepler, others continued and enhanced the work they started.[38]

It was this period that created the classical image of scientists as being purely objective in their investigations and that the theories of science were equally objective. This notion, of course, was never completely true. Even though the scientists were not considered part of experimental processes, there was always a personal aspect of investigations based on the scientist's inclinations, objectives, and agendas (possibly hidden). Therefore, even though mental aspects were not considered part of science and, while the experiments themselves may have been objective, the conjectures and hypotheses based on these experiments necessarily contain subjective elements influenced by the scientist's personality.[39]

In any case, a major concern of classical science was force acting on matter, also known as dynamics. Two remarkable individuals initiated this trend. The first was the above-mentioned Galileo who pioneered modern scientific method through the innovation of combining experiment

---

[38] The reader may have noticed that it is traditional to refer to scientists and other famous people by their family (last) name with one exception – Galileo is always referred to by his given (first) name. How this came about is another mystery to be solved.

[39] Herein, the *objective* is considered a reference to something independent of a perceiving mind, while the *subjective* is somewhat the opposite and refers to some notion held in the thinking mind. Put another way, objective is what we believe as having independent reality and subjective is a personalized interpretation.

with mathematics. The second was Isaac Newton (English Physicist, 1642-1727) who, utilizing the kinematic equations of Kepler, brought gravity into the study of planetary motion.

Although Newton wrote extensively about gravity, he never attempted to determine its source, or at least, did not publish any such attempt and it was not until the 19th century that gravity began to be recognized as an intrinsic property of matter and it was conjectured that gravity is a *field* propagated by some unknown mechanism. Put simply, in the scientific sense, a field is something that propagates between objects and has variable influence on these objects [See *Side Comment on Fields* below].

After gravity, the second fundamental field discovered was the electromagnetic field (EM Field). This was the work of James Clerk Maxwell (English, 1831-1879), who built on the earlier work of Johann Carl Friedrich Gauss (German, 1777-1855), Michael Faraday (English, 1791-1867), and others. In a manner, analogous to how Newton's dynamics supervened on Kepler's kinematics, Maxwell's electrodynamics supervened on Faraday's electrostatics.[40]

Then in 1896, just as everyone was getting comfortable with notions about electricity and EMF, along came the discovery of radioactivity. This

---

[40] Most forces that we know about have some theoretical propagator associated with them, even if this propagator has yet to be *discovered*.

was especially surprising and disturbing, because at that time the concept of the atom, although still being debated, was thought to be indivisible. The discovery of radioactivity settled the debate about atoms, but more importantly, it radically changed the notion of an atom from something monolithic and indivisible into something with parts, and sometimes one or more of these parts are *kicked out* of the atom!

Side Comment on Fields: A semi-technical definition of a field goes something like this; "a field is a structure of continuously varying functions of position and time that are usually described mathematically using differential equations."[41] Or, as Richard Feynman (American Physicist, 1918–1988) put it; "Physicists use the word 'field' to describe a quantity that depends on position is space and time."[42] The most commonly encountered fields are the electromagnetic field (EM Field) and the gravitational. The EM Field is conjectured to be propagated by microscopic bundles of energy called photons. On the other hand, it has been proposed that the gravitational field is propagated by a particle called a graviton, which, if it exists, has eluded all attempts at discovery. It was the 19th century that saw great progress in examining and understanding what eventually came to be called the electromagnetic field. As the term implies this field is a relation

---

[41] Wolfgang Pauli, *The Theory of Relativity* [Supplemental Note 23] (New York: Dover Publications Inc., 1981), 142.

[42] Richard Feynman, *QED: The Strange Story of Light and Magnetism.* (Princeton University Press, Princeton, NJ, 1985), 123.

between a magnetic field and an electric field such that each can induce the other through relative motion. The EM field could be considered the third field since the magnetic field has been known since antiquity – although it was not until the 19th century that it was recognized that magnetism was associated with the electric field.

As a magnet moves relative to a closed loop of metal somehow the magnetic field induces an electric current as electrons[43] in the metal loop move in a certain direction and in quantities proportional to the strength of the magnet, while on the other hand, an electric current such as that produced by connecting a wire to a battery causes electrons moving through the wires to induce a magnet field.

There seems to be an obvious complementarity between electric charges and magnetic fields that are understood primarily only by the observed behavior and not through great insight into the actual causes of this relationship. There can be no rational doubt that the rotating magnetic field in a generator reaches out through space to induce electrical current but even after being studied and in practical use for generations we still do not really know why this actually happens. That is, we do not know exactly what causes a magnetic field except that it seems to have something to do with the inherent magnetic tendencies within certain metallic atoms.

---

[43] Electrons are elementary particles and are one of the basic building blocks of atoms. They are known to absorb and eject photons, as well as generate an electro-static field when motionless or an electromagnetic field when in motion.

In keeping with the propensity to attribute fields to specific field generating particles, propagators of the magnetic field have been proposed along the same lines as the postulated photons of electromagnetic radiation (aka light); however, the conjectured propagator of magnetism has not been found. It is just another one of the great mysteries of nature that for all our assumed intellectual capacity cannot – at this time – solve. However, since both a magnetic field and the electromagnetic radiation can propagate through vacuums it would seem that *some-thing* must be passing through these vacuums.

Sided Comment on Dynamics and Kinematics: The differences between kinematic and dynamic motion are examples of how one theory can supervene on another. Kinematics is essentially a special (simplified) version of dynamics where the equations of motion are geometric trajectories and any forces that might be acting on the object of study is ignored, while in dynamics, the equations of motion include not just the geometric trajectories but include the forces acting on the object and the inertia of the object. Put simply, kinematics is about non-causal motion, while dynamics offers causal explanations for this motion.

Side Comment of Scientific Definitions: Science is a system based on definitions. That is, all propositions of a scientific theory are based on terms, which have been defined by other terms. At some point, there must be unproven terms and propositions that are accepted without proofs, since this is the only way to avoid either an unending series of definitions or circular definitions. This essentially means that not all propositions within a scientific system can be proven and this is one reason that theories are always subject to replacement or revision. Moreover, scientific endeavors are restricted to the unique set of notions that govern activities within each discipline (physics, chemistry, biology, …). In this way, each discipline is a closed system within which the scientist must work. For example, a chemist 'sees' the structure of matter differently from that of a physicist.

Side Comment on Coordinate systems: Science and engineering have long relied on the coordinate system devised about 400 years ago by the philosopher and mathematician Rene Descartes (French, 1596–1650). In two-dimensions, this is the familiar system based on two perpendicular lines generally referred to as the x-axis and the y-axis. Although this technique of using these XY-coordinates for science and engineering has been highly effective and useful, it represents a rather confined way of thinking. The problem has always been how to replace this system and with what?

The answer to this may be both somewhat old and somewhat new. Around the middle of the 19th century, an English mathematician, William Clifford (1845-1879), came up with an idea for a new algebra, but unfortunately, Clifford died before he could fully develop his new algebra. Consequently, his idea remained essentially dormant for about a hundred years until a physicist named David Hestenes (American, 1933-) rediscovered Clifford's work and pioneered a new geometry called Geometric Algebra, which has its beginnings with Clifford's algebra. This Geometric Algebra is essentially coordinate system independent, which makes it adaptable to many types of problems. Although this algebra has not replaced the old-fashioned Cartesian coordinate system, it is making gains and appears to have a bright future. By the way, you know all those dazzling special effects that you see in the cinema, geometric algebra plays a key role in making that possible.

Side Comment on Time and Space Relations: Regarding space and the location of objects, it must be recognized that there can be no reference to a *there* without a *here*. Likewise, regarding time, there is no *later* without an *earlier*. In the case of the elusive *now*, we sandwich that in between the past (memory) and the future (anticipated *now*). That is, both space and time are determined by relational concepts relative to an observer's position and all spatial and temporal statements imply or explicitly state a comparison (there-here, before-after).

Scientific Revolutions of the 20th Century

At the beginning of the 20th century, two revolutionary events occurred in science. The first was the development of the theories of relativity that forced a complete reevaluation of our classical notions of space and time. The second was the development of quantum theory, which was even more revolutionary than relativity. Quantum theory, as it developed, revealed that the behavior of matter at microscopic levels was stranger than could have been imagined.

The first revolution started in 1905 when Albert Einstein (German Physicist, 1879-1955) published a paper that eventually led to the special theory of relativity. At first, this paper did not receive much attention and he did not become an overnight sensation. However, after several influential physicists read, and understood his paper, its significance began to be recognized. However, even then, it was still several years before his theories gained wide acceptance and brought Einstein worldwide fame.[44]

One of the few who, early on, recognized the significance of relativity was the mathematician Hermann Minkowski (Polish-German, 1864-

---

[44] It is worth noting that in 1905, not only did Einstein publish the paper on special relativity, but he also published three other papers, any one of which would have made his reputation as a first-class physicist. Moreover, one of these was a seminal paper for the future quantum theory and the one that later earned him a Nobel Prize.

1909), who, in 1908, made a speech where he pronounced that relativity theory connected space and time such that he coined the term *space-time* (or spacetime) to indicate this bonding.[45] That is, Minkowski proposed that time and space should be considered inextricably connected both mathematically and physically. In addition, to this paradigm change for space and time, the special theory introduced a famous equation that proclaimed the equivalence and mutual mutability of matter and energy. Several decades later, in 1945, the reality of this equivalence was dramatically demonstrated with the explosion of the first atomic bombs.

Ten years later, in 1915, Einstein, being no one-hit wonder, was again able to astonish the scientific world with the publication of his generalization of the special theory of relativity. As special relativity changed the status of time, general relativity changed the status of space. Put simply, special relativity specifies the speed of light as a fundamental constant of nature and demoted time to a variable (more about this later). On the other hand, general relativity specifies gravity as fundamental and space became subordinate to gravity. Within the *standard interpretation* of general relativity, gravity determines what kind of space exists. Instead of the flat Euclidean space,

---

[45] Minkowski announced this proposal in a Lecture delivered before the *Congress of Natural Philosophers at Cologne* (21 September 1908). Unfortunately, Minkowski died just months later and was not able to complete the development of his idea.

space is assumed to be curved under the influence of gravity – later we will reconsider this notion about space.[46]

* * *

Side Comment on Relativity: Put simply, the basic difference between special relativity and general relativity is the introduction of gravity, which is missing completely from the special theory. In special relativity, forces that move objects and the momentum of objects are not included and this is one of the motivations for developing the general theory, which included both and more. The general theory also includes stress and tension with the result that the mathematics of general theory are considerably more complex than that found in special theory. This is one reason why it took Einstein so long to develop the general theory, he had to learn new mathematics.

* * *

Although the theories of relativity introduced radical changes in our understanding of space and time, it still maintained many of the characteristics of classical science. For example, within both relativity and classical science, nature was considered deterministic, causal, reversible, and continuous. By retaining these classical notions, relativity may be considered the end product of classical physics, while quantum physics is the

---

[46] Euclidean space is named after the fourth century BC Greek mathematician, *Euclid of Alexandria*, whose work called *The Elements* has essentially been in print for more than 2000 years! Euclidean geometry of space is analogous to flat surface such as a basketball court.

beginning of a new physics that no one had imagined possible.

Quantum theory indicated that at the micro-level, far removed from direct human experience, the behavior of matter was just the opposite from classical science. That is, quantum theory was indeterminate, acausal, irreversible, and discrete! Moreover, it seems that quantum theory has a mental element never encountered in science. While classical science used terms such as velocity, force, and momentum, quantum theory introduced into the lexicon of science strange unfamiliar and sometimes vaguely defined terms such as, uncertainty, entanglement, complementarity, duality, and exclusion.

Put simply, a large part of classical physics was about the behavior of matter at the macro-level, while quantum theory is about discrete states of matter at the micro-level (atomic or sub-atomic level). That is, classical mechanics is about smoothly connected behavior of extended matter found in both kinematics and dynamics. This is why the differential calculus, which has the appearance of unbroken connectedness, is so useful in classical science. On the other hand, in quantum theory, things do not flow; things apparently change states incrementally, without passing through any intermediate states.

For example, the classical description of a rock thrown through a glass window is an event in which the rock has a specific trajectory that passes through the window and results in a broken window. On the other hand, the quantum version would be a description of the state of the rock and the state of the window prior to the window being broken and the state of the rock and the state of the window after the window is broken.[47] There is no specific trajectory and there is no direct evidence that the rock broke the window, there is only the implication or probability.

Classical events have direct causality, while quantum events seem to have no direct causality. There are only before and after conditions or states. That is, there is no proof that the rock broke the window; there exists only a probability that the rock broke the window. Or, as stated by Gary Zukav, there is a *tendency* for certain things to happen and exist, which are expressed as probabilities.[48]

The replacement of the classical scientific notion of cause and effect by a *statistical correspondence* is a necessary adjunct to the study of micro-phenomena. The gap between observer-interpreter (a human) and the microscopic physical system being examined is mediated through many levels of extrapolation, amplification, and simulation such that the results are not only statistical, but sometimes highly speculative.

---

[47] Physics uses the term state to describe possible results or configurations. For example, a coin toss results in either of two states – heads or tails, while the toss of a die has six possible states.

[48] Gary Zukav, *The Dancing Wu Li Masters*, (New York, NY, Perennial Classics, 2001), 35.

The uncertain nature of quantum theory requires the introduction of mental elements as a means of bridging the gaps in the quantum level behavior of matter. Quantum events have a three-part aspect consisting of a mathematical description, a physical description, and a mental description. The mathematical (behavioral description) are in the equations of the theory. The physical descriptions are in the experimental setup and apparatus, while the mental elements are in the choices of what to measure and the interpretation of the instrument readings.

For example, the experimenter must make conscious choices about what will be measured. These choices include such things as measurements of either position or momentum (but not both), and wave-like or particle-like behavior (but not both). In this sense, the experimenter is no longer merely the *observer* of classical science but has become a participator.

Despite all the strangeness or, maybe, because of all the strangeness, the founders of quantum theory, for the most part, were not prone to examine or ask fundamental questions about the theory. This attitude was mainly due to Niels Bohr (Danish physicist, 1885-1962), who, as default leader, shunned fundamental thinking about the theory. Bohr held that quantum theory was a *statistically driven operational theory* and that there was no good reason to seek a more fundamental understanding. With some exceptions, this attitude has continued to the present.

As quantum theory developed, it became obvious that there comes a point in the study and contemplation of the natural world where the classical notions of matter, space-time, and causality no longer applied and a new (or perhaps very old) way of thinking is necessary. In between the determinism of classical science and the apparent randomness of quantum phenomena, there exists the unpredictable and quasi-determinant middle ground where the randomness gradually gives way to determinism. That is, phenomena such as the famous *complementarity*, which is related to wave-particle duality slowly gives way to behavior that begins to resemble the predictability of classical science (See the *Side Comment on Complementarity* below).

There is speculation that this strange quantum behavior never completely goes away, but simply the probability diminishes asymptotically – in the sense of approaching but never quite reaching zero - as matter scales from the micro-levels to the macro-levels. This would mean that, although the probability would be extremely low, something strange like entanglement (See the *Side Comment on Entanglement* below) could occur at the level of human experience. For example, sudden and unexplained incidents might occur, such as those associated with synchronicity (to be discussed later).

Exactly where this transition takes place is something of an open question. The wave-particle duality of quantum behavior has been observed in molecules of up to 60 carbon atoms, which is immensely greater in mass than a photon[49] or an electron,[50] but is still far below the macro-level of everyday experience. When we speak of the wave-particle duality, it should not be considered that observation of this duality means that the electrons or photons are actual particles or waves. That is, electrons and photons are neither waves nor particles, they are something else that only seems to behave like either a wave or a particle – depending on the experimental setup.

It is interesting to note that in 1906, JJ Thomson won a Nobel Prize by demonstrating that electrons are particles, and in 1937, his son, George Thomson won a Nobel Prize by demonstrating that the electron was a wave. Looks like you can have it both ways – so, take your pick.

When thinking about *waves*, we tend to visualize them as having crests and troughs such as those observed on the surface of a body of water. This is OK, up to a point, but when contemplating the dual nature of particles, this description is totally inadequate. Wave-like behavior does not mean,

---

[49] Photon is the name given to the conjectured propagators of light, also known as electromagnetic radiation. These things, conjectured to be miniscule blobs of energy, only travel at the speed of light and essentially do not exist other than when they are traveling at the speed of light.

[50] Electrons are elementary particles and are one of the basic building blocks of atoms. They are known to absorb and eject photons, as well as generate an electro-static field when motionless or an electromagnetic field when in motion.

in this case, an actually wave. This oscillatory behavior of elementary matter is a mathematical
description based on observed behavior, not necessarily a physical behavior. What it really is, we
do not know, we only know that these particles
exhibit properties that are inconsistent with solid
matter, as we understand it.

Side Comment on Seeing the Light: In his book on
quantum electrodynamics Richard Feynman insists
that photons are particles.[51] This is in spite of the well-
known wave-like behavior of photons as experienced
in some experimental setups. However, given the discrete nature that is postulated for matter at the quantum
level and the fundamental equivalence of energy and
matter, it is reasonable to subscribe to the belief in the
particle-like (or packet-like) nature of light (photons).

This, however, does not mean that light propagators
are something like a grain of sand or speck of dust,
which has definite size and shape. Photons and other
micro-level particles have a different nature that allows them to manifest both wave-like and particle-like
properties while being neither. We call them particles
because we have no better word available.[52]

On the other hand, Feynman was incorrect when he
said that quantum electrodynamic explains all of
chemistry. For example, it explains neither chirality
(handedness such as the fact that nature prefers left-

---

[51] Feynman, 1985, 14, 37.

[52] At one time, the word *wavicle* was proposed, but it never
caught on.

handed molecules) in molecules nor does it explain chemical properties.

Side Comment on Quantum Jumps and Quantum Leaps: One of the most common misconceptions and misuses about quantum theory are references to jumps and leaps. Scientists sometimes use the term quantum jump (or leap) when referring informally to the transfer, within an atom, of an electron from one energy level to another. Unfortunately, one often hears the expressions '*quantum leap*' and '*quantum jump*' used in the sense of a significant or notable change or enhancement to some thing or process. This is a strange and inverted usage of what is perhaps the smallest increment change within nature, and it is very unlikely that anyone who actually knows something about quantum theory would ever use this expression in such a loose manner.

Side Comment on Complementarity (and wave-particle duality): It seems that we attribute forces like electromagnetism to phantom-like elementary particles called photons. These *particles* far too small for direct observation, have an existence that is an inferential extrapolation derived from the measurement of various phenomena arising out of experimental processes. However, these particles play tricks on us by changing their behavior. Perhaps the most famous and one of the simplest examples of this schizophrenic or bipolar behavior is the so-called *two-slit* experiment.

This experiment is conceptually quite simple, consisting only of a light source, an opaque screen, and a target such as a strip of film. First, two very narrow parallel slits are cut close together in the screen and then the screen is placed between the light source and the target. If one of the slits is covered and light is projected through the open slit onto the target, a solid bar of light appears on the target.

However, if both slits are unblocked and the light is projected through both slits the expected result is two light bars each similar to the one observed when only one slit was open. However, instead of the expected pattern, a wave pattern appears on the target [alternating light and dark bars]. What this indicates is that, with only one slit open, the behavior of the light is particle-like, while with both slits open, the behavior is wave-like. This means that you can measure either wave-like or particle-like behavior, but not both at the same time. This behavior is known as complementarity.

Do not try to figure it out; people have been trying, without success, for more than two hundred years! Thomas Young (English polymath, 1773–1829) first performed this experiment in 1803.

Side Comment on Entanglement: Entanglement is a condition where two (or more) particles such as electrons or photons become linked through some unknown process that binds them in a manner that is called entanglement. The result is that an action on one of these particles is instantaneously *felt* by the other. Essentially, the two particles act as if they were one particle. Additionally, this condition remains no matter how far apart the particles are separated. That is, spatial separation of any distance has no effect on this condition. This condition is another example of how quantum theory teases us by hinting at a reality that is yet to be discovered.

Side Comment on What is Energy?: First, we know this is another one of those words that get used commonly in such expressions as *Today I feel low on energy* or to remark that someone is *energetic*. This is fine but we also know that energy has something to do with fuel or caloric intake such as sugar. If we were to get somewhat technical, we would find that there are different kinds of energy.

Here is a list of nine types of energy: *kinetic* energy, *gravitational* energy, *electrical* energy, *chemical* energy, *radiant* energy, *nuclear* energy, *heat* energy, *elastic* energy, and - last but not least – *mass* energy. Most of these are familiar, for example, batteries use chemical energy to produce electric energy, radiant energy from the sun fuels life on this planet, nuclear energy is used at nuclear power plants, and mass energy is the source of energy for nuclear bombs.

We know a lot about each of these different forms of energy and they all provide capabilities that are essential for our world to function the way it does. Within science, technology, and engineering we use various mathematical formulas to calculate these numerical quantities we call energy. But, after centuries of learning about these various kinds of energy we do not really know what it is - so, I simply cannot answer the above question.

Scientific Theories of Today

"…The research on the matching of general relativity with the quantum nature of matter is still in its infancy; however, we already understand that reconciling these two pillars of the XX's century physics will require a whole new set of ideas, as it may happen, that we may have to give up concepts that were supposed to be the starting point of all the modeling of the universe, such as that space-time is a continuum and that the fundamental building blocks of matter are not point-like particles."[53]

"It is not always possible to check by experiment every separate assertion of a scientific theory, although the system of thought as a whole must, if it deserves the name of a scientific theory, contain possibilities of a check by empirical methods. This is what constitutes its verifiability."[54]

First, what is a theory? The dictionary uses words such as a supposition or a set of suppositions that have the intent to explain something. Other closely related words include hypothesis, conjecture, and speculation. However, instead of these words, the word theory is used a lot within the scientific community because it apparently seems more respectable and is considered to have a

---

[53] Orfeu Bertolam (2008). *The Mystical Formula and the Mystery of Khronos* (<arXiv: gr-qc/0801.3994v1>).

[54] Wolfgang Pauli, *Writings of Physics and Philosophy*, Eds. C. Eng and K. von Meyenn. (Springer-Verlag: Berlin, 1994), 139.

meaning closer to *fact* than the other words just listed.

Nevertheless, no matter how many times a theory is correct, there is always the possibility that it might require modification or even revocation. For example, the special theory of relativity, first published in 1905, has passed ever test of its validity for more than a hundred years, but is still ranked as a theory and has not been promoted to a *first principle* or *law*.[55] Put simply, a scientific theory includes, at least, these four things: a theoretical interpretation, a consistent mathematical formulation, the possibility of experimental confirmation (or the possibility of falsification), and prediction of previously unknown phenomena (that is, no hand fixing to match previously known observations). If any one of these things is missing then it is a stretch to call something, a *scientific* theory.

Put simply, to accept a theory is take a speculative chance. There are always alternative choices, and the choice of selection is not always about better or worse theories, but rather about acceptability within certain established guidelines or intersubjective agreement based on personal

---

[55] A first principle is a proposition that cannot be deduced from any other proposition, postulate, or conjecture. A scientific law is based on persistence of observations of the workings of the universe where no explanation is given, but there is an assumed causality.

and/or social agendas as well as the possibility of groupthink or other means of peer pressure.

There is a tradition that states, all things being equal, the simpler theory is, or should be, the theory of choice. In practice, this is not always a correct method of selection. Simplicity alone does not suffice as a valid reason, there must be other considerations such as refutability, descriptive accuracy, evidential support, completeness, consistency, novel predictions, and so on.

For example, the theories of relativity are definitely more complex than the Newtonian theories that they superseded. Another example was the replacement of the geo-centric concept of the solar system with a helio-centric configuration was not a simplification in that unlike the old model, where the earth was stationary, the new one requires that the earth have both an annual orbit around the sun and a daily axial rotation.

Both of these examples profoundly changed the way we understand the natural world and superseded earlier theories. However, it seems that now there is a trend to patch-up theories, rather than replace the old or untenable theories with new theories. Additionally, not all of these patches are motivated by valid scientific reasoning but are made to save theories for non-scientific agendas.

Theories are not proven through repetitive observations or experiments; they are only made more probable. That is, inductive inference and consistency cannot prove a theory, it can only increase the probability of truth and increase confidence in the theory. On the other hand, one negative result can damage or even invalidate a theory.

It has been claimed that a necessary criterion of demarcation between science and metaphysics (or some pseudo-science) is its refutability. For example, according to Karl Popper, (Austrian-British philosopher, 1902 – 1994), a system may be considered science only if it has been or can be tested by attempts to refute. As Popper said, "A theory which is not refutable by any conceivable event is non-scientific. Irrefutability is not a virtue of a theory, but a vice."[56] Many consider this position too extreme, but, on the other hand it seems intuitively apparent that there needs to be some objective criteria for theory validation.

Popper chose falsification for two main reasons, first, scientific theories such as Newton's *Laws* can never be fully verified but can be invalidated by a single false example. Secondly, by choosing falsification, Popper setup a demarcation between statements that can and those that cannot be found false. For example, in today's Standard Model of Cosmology, inflation is certainly not a falsifiable notion, and, in this case, it is also not a verifiable notion. As such, cosmic inflation should not be considered a true scientific theory

[56] Karl Popper, *Conjectures and Refutations: The Growth of Knowledge*, (New York, Harper and Roe, 1963), 36.

but merely a speculative notion created to mend the big-bang conjecture [See the section *Multiverse and the Fine-Tuning Argument* - page 142].

True science is characterized by its observational and analytical methods, while pseudo-science (or metaphysics) is characterized by its speculative method (or mental anticipations?) Irrefutability arguments may use either logical or empirical methods (or both?). An example of a *logical* irrefutability are the statements '*It is raining*' and '*It is not raining.*' These statements are logically irrefutable, although one or the other must be false. An example of *empirical* irrefutability is a statement of existential fact, such as '*I own a pair of brown shoes.*' Irrefutable does not necessarily mean that a theory is true. For example, philosophical theories are irrefutable, but they may also be false.

On the other hand, these days scientific theories fall into two broad categories; those that are empirical and those that are non-empirical in the sense that they are for the most part, if not entirely, observational (such as astronomy). Truly empirical theories are refutable, while the non-empirical are irrefutable for much the same reason that philosophical theories are irrefutable. Philosophical theories may be argued against, but falsification is not possible. That is, irrefutable in the sense of not being able to prove wrong, but, of course, are still deniable.

*   *   *

Science is obviously a human endeavor, but less obvious are the limitations or phenomenal boundaries within which scientific efforts are pursued such that the scientist is restricted to phenomenal information only within these boundaries. That is, each scientific discipline has their own boundaries that define domains of validity and discourse. For example, a chemist has ways of dealing with chemical elements and the structure of molecules that differ from that of a physicist. This means that if the physics community developed something that seemed a valid theory of everything, it would not necessarily be a theory of everything for a chemist, and, therefore, not really a *theory of everything*.

Scientists must necessarily restrict deliberation and investigations to only a simplified version of nature whose complexity is too great for total consideration by any means currently available. That is, scientific studies are by necessity restricted to that which can be measured physically and analyzed mathematically. The complexity of reality as experienced for the most part falls outside the narrow domain of the scientist such that beauty, the mystical, the psyche, and so on fall outside of the scientific domain.

There is an element of arbitrariness in scientific investigations because the scientist must make a choice of what to examine and what to ignore such that the real world as examined by a scientist is always a selective simplified version of reality [the reader will recall the Oppenheimer quote on page xiii]. This not to say that this technique wrong, quite the contrary this technique has over the years proven itself to be very successful. In fact, it has been so successful that scientist began to believe that they had captured the real world behind the world of phenomenal reality.

Of course, all these scientific discoveries and investigations ignored and continue to ignore the realities of value and intuition which lie outside of the scientific domain such that the mystical, transcendent, beauty. Put simply, science and scientists are sanctioned only to deal with the metrical, the qualitative and the quantitative of the phenomenal world, are not equipped to deal with these other realities.

Lacking such virtues and values, scientists – under the influence of declining religious convictions – have adopted materialism and the philosophy of meaninglessness. That is, lacking purpose or value they embrace a mechanistic view of life that reduces live creatures to mere machines.

Because of their specialized training, scientists, for the most part, seem to be neglectful of values beyond their chosen profession such that they have not been able to satisfy the natural human striving for that which is greater, and have attempted to satisfy this hunger through social and political agendas that are progressively liberal. Since many scientists today profess to be atheists and deny the existence of a soul, they seek their religious conation through extreme and exaggerated social agendas such as global warming where they have adopted the *chicken little syndrome*.

True and honest science exists only where there is open-mindedness and truthfulness along with a willingness to consider alternatives. It is worth repeating that the accumulation of evidence does not prove a theory, it only makes it more plausible or more likely to be true. This is not to say that theories are falsehoods or make-believe, rather they are only simplifications and approximations of reality behind perception and this is likely to always be the case.

*   *   *

As mentioned in the first quote at the start of this section, along with the notion of a theory of everything, there is the current inability to unify the general theory of relativity (gravity) and quantum theory (micro-level behavior of matter) is an example of how physics has almost come to a halt. On the other hand, given the contradictory nature of relativity theory and quantum theory, it is not surprising that all attempts to join these two theories have failed and will likely continue to fail unless some novel approach to the solution is found, which will likely fall outside of traditional mechanistic thinking.

It seems that the glory days of science are over – at least temporarily – and except for life sciences pure science seems to have made little real progress since the end of World War II, which, more than any other thing changed pure research science from what was essentially an individual effort into BIG science such as that which produced the first atomic weapons.

Individual research and the pursuit of pure science is a very difficult task that requires more than just intelligence, it also requires years of hard work and a perseverance to do this hard work. The problem is that few seem willing to do this or are the remarkable scientific revolutions of the early 20th century just too difficult to improve on in any real way. Or, perhaps that as science in general has grown through the increasing necessity to specialize that the big picture that led to the past great discoveries is just too difficult to wholly grasp?

It is interesting to note, that both relativity theory and quantum theory are more than one hundred years old but there is still no major advancement of either these theories or some new theory that might have a similar impact as these older ones. OK, in all fairness, it should be mentioned that the theories of Newton lasted almost three hundred years before they were rendered incomplete by the theories of relativity.

However, today there are thousands of physicists working on some aspect of physics today which is far more than were active worldwide a hundred years ago. It should be recalled that the theories of relativity, as originally published, were essentially the work of a single person – building on the previous work of others, and the building of quantum theory, primarily in the 1920s, was the work of a handful of inspired and dedicated scientists. Since then progress in fundamental physics has, with a few exceptions such as quantum electrodynamics – also the work of only a few, been all but stalled.

Whether this lack of new game changing theories is because of the lack of insight or whether the pursuit of science has fundamentally changed such that *seeing* beyond these theories is just too difficult. It is a fact, that massive projects like the LHC in Switzerland has provided employment for literally thousands of scientists and engineers, has cost billions of dollars, and to what end? Perhaps big science or science by committee is not capable of the type of insight that occurred up to the middle of the 20th century.

In any case, it seems that many of the so-called advances that are touted in the press are not particularly reliable with regard to being really scientific or are they just pipedreams of wishful thinking masquerading as true science?

Perhaps, physics took a wrong turn and became lost in theories and conjectures that have led nowhere? Some have even speculated that all attempts to merge relativity and quantum theory are doomed to failure because of incompleteness and/or inconsistencies in both theories. Physics, in general, and the Standard Model of Particle

Physics,[57] in particular, seems stymied and at a stalemate.

Here is a short list of some current issues:

1) String theory and M-theory came and went.
2) Quantum gravity seems to be going nowhere.
3) Unverifiable conjectures such as the *multiverse* have gained in popularity.
4) Particle physics seems a self-fulfilling prophecy.
5) Unquestioned adherence to the big bang and cosmic inflation conjectures.
6) Unquestioned adherence to the Standard Model of Particle Physics.

These are just a few of the many issues in physics and fields relate to physics. If we were to venture into biology, chemistry, and so on, we would find that these fields also have many unsolved and, perhaps, unsolvable problems.[58]

*   *   *

The empirical knowledge of science is always subject to revision due to some future insight, measurement, or observation. The scientist deals

---

[57] The Standard Model of Particle Physics is a reference to current conjectures about elementary particles and nuclear forces. Although, this model contains numerous discrepancies, most notably, it does not include gravity.

[58] For those who might be interested, Wikipedia has lists of unsolved problems for physics, chemistry, biology, and more.

with phenomena and must always be aware that beyond the phenomena is the unknowable thing-in-itself. Given this, it is not surprising that science uses conditional words such as theory, postulate, hypothesis, and conjecture when dealing with its subject matter. There is always an element of incompleteness and uncertainty.

The Standard Model of Particle Physics is a prime example of what has been sometimes called an infectious idea that grows in both content and complexity in pursuit of some final conjecture. This is why the number of so-called elementary particles exploded out of control and has created a culture with an inextricable particle mentality. That is, the true believers simply cannot conceive of any solution that is not based on elementary particles.

As stated previously, no current scientific theory will escape modification or possibly even revocation and replacement. All knowledge is in flux and as we learn more about our place in the universe even religion is not exempt from change and reinterpretation. In any case, up to the present, efforts to develop a theory of everything or a Grand Unified Theory, as it is sometimes called, have not been successful and these efforts seem likely to remain unsuccessful.

Side Comment on Particle Physics and the CERN LHC: CERN is the European Center for Nuclear Research located in Geneva, Switzerland. This facility includes the Large Hadron Collider (more commonly called the LHC), which is the world's largest and most powerful sub-atomic particle accelerator. The LHC is a massive group of machines, that includes a circular tunnel 27 kilometers in circumference. Its purpose is to investigate the nature of matter at microscopic levels, well below the size of atoms.

One of the main justifications for this multibillion-dollar operation was expressed as the search for a particle known as the Higgs, which is named after the physicist who first proposed the possibility of its existence. To be sure, about the time that the LHC went operational in 2008, there was no doubt in my mind that the Higgs particle would be found. *This is not because I believed in the existence of the Higgs particle but because they had to find it. This was the only way they could justify the great effort and massive expenditures*.

Sure enough, after a few years of operation, they eventually claimed to have *discovered* the Higgs and this *discovery* was officially published in the 2012 edition of the *Review of Particle Physics*. This document contains 1526 pages and mentions the Higgs, 1122 times, but what is more interesting is that the word *assume* (or variations on this word such as assumption) is used 522 times![59] Obviously, the Higgs is the star of the show, but the use of so many assumptions seems to imply a lot of guesswork rather than hard evidence.

By the way, the 2016 version expanded to 1808 pages, mentions the Higgs 1823 times, and uses the word

---

[59] The 2012 edition of *Review of Particle Physics* is published for the Particle Data Group in Phys. Rev. D86, 010001 (2012) [http://pdg.lbl.gov/]

assume or one of its variations 723 times![60] The recent 2018 version (1898 pages) seems to have leveled off somewhat since the Higgs was only mentioned only 1740 times and the word assume or variations was slightly down at 712 times.[61] The fun never stops!

Side comment on Creativity: Consider these three words; discovery, invention, and creation. They represent three different ways we advance knowledge and all three have very different meanings and reference very different activities. Discovery is what explorers and archeologists do. They find things and then try to attach some significance or meaning to what they found. On the other hand, inventions, which might be described as the assembly of known parts into some novel thing. This is what engineers, tinkerers and, possibly, scientists do. At the peak of this hierarchy of these three words, is creation. True creation is, among other things, an act of seeing things differently. Creation is not about quality or even esthetics, but it is about uniqueness and not utility or application. It is producing something unique and is more than invention, as invention is more than discovery.

---

[60] The 2016 edition of *Review of Particle Physics* is published for the Particle Data Group in Chinese Physics C Vol. 40, No. 10 (2016) 100001 [http://pdg.lbl.gov/]

[61] The 2018 edition of *Review of Particle Physics* is published for the Particle Data Group in Phys. Rev. https://journals.aps.org/prd/abstract/10.1103/PhysRevD.98.030001

Scientific Dogmatism

"The essence of scientific freedom is the right to come to conclusions which differ from those of the majority." [62]

Scientific dogmatism seems to be more prevalent today than it was the past. At the beginning of modern science, science had to be careful about conforming to the dogma of the western church. Today, instead of religious dogma, a scientist is required to follow the prevailing scientific dogma. In the 16th century, at the beginning of modern science, one could lose one's life, today one can remain unemployed [See *Side Comment on Giordano Bruno* following this section]. The punishment is different, but the result is the same, innovation and freedom of thought is stifled.

It is the goal of all scientific endeavors to determine objective characteristics of the natural world. However, at the same time, science is also trying to undo, modify, or extend these previously determined objective characteristics. The result is at best a complementary situation, but, in practice, it is more like a conundrum or contradiction. There exists a sort of *push-pull* effort between the desire to discover the new and the desire to preserve the old. Failure to entertain novel ideas or the possibility of change leads to

---

[62] Edward Arthur Milne, *Modern Cosmology & the Christian Idea of God* (Edward Cadbury Lectures, 1950): (Oxford: Clarendon Press, 1951), 8.

stagnation and dogmatism, while readiness to consider novel ideas leads to innovation and true progress. Of course, this readiness should not be hasty acceptance of notions with little supporting evidence.

Science has moved from an understanding of its limitations and what it actually states, to a more socially motivated position where the limitations of knowledge about the natural world are encased in elaborate theories that are proclaimed to the world as fact and the truth about nature even though many of these statements rest of weak foundations or are simply made-up ad hoc statements used as patches on some grand conjecture that continues to grow like some *Tower of Babel*.

For example, the latest (2019) example is the proclamations about the discovery of remote planetary systems which now total into the hundreds – this is reminiscent of the proliferation of so-called elementary particles that grew with each increase in accelerator power and now number in the hundreds. I might add a word of clarification. I do not doubt that billions of other planets are out there, I just recommend caution concerning these *discoveries*. These findings should be considered preliminary and not a proven scientific fact. Error rates have been estimated as high as 35%, which is not far from guess work.

Without doubt, reluctance to consider novel ideas has become dogmatically engrained within the scientific community. Those who dare question the orthodox risk being *burned at the stake* of rejection and ridicule! Unfortunately, this is not an entirely new thing, as far back as 1927, at a conference in Copenhagen, Louis de Broglie, an eminent French physicist, was treated rather badly by several of his colleagues. De Broglie had made a proposal that did not fit the current quantum theory dogma as dictated by the inner circle at Copenhagen. Only recently have his ideas been given a fair hearing.

The community of physics orthodoxy is full of mathematically competent practitioners who scorn and ostracize dissenters, it seems that there is no longer any place within this culture for free thinkers of the type that dominated physics until the middle of the 20th century. These innovative physicists of the first half of the last century set a very high bar for hard work, creativity, and accomplishment such that few if any remain of that caliber. Today, as a result of big science and the so-called new physics, today's technocrats bow down to half-baked conjectures that seem to pop up without justification other than the belief that they need to look like they are making progress.

Side Comment on Big Science: Over the last few centuries, as the scope of scientific knowledge grew, it began to split into ever more specific areas of specialization, which caused a fragmentation and isolation between these specialized areas. To be sure, the vast scope of scientific knowledge makes it impossible for any individual to grasp much more than their chosen discipline. Consequently, the age of the 'polymath' or the multi-disciplined natural scientist is essentially over. Complexity is reduced by dividing large problems into smaller chunks that can be manipulated and in this way problems that otherwise might prove too difficult are successfully solved. The down side is that this results in even more fragmentation that makes it easy and even likely that one will lose sight of the connections that link these scientific divisions.[63] As a result of this increase in narrowness of focus, the need for scientific collaborations has increased greatly such that now there are consortiums that employee hundreds (or even thousands – such as the above mentioned CERN LHC) of scientists and engineers in pursuit of solving problems and understanding nature.

Side Comment on Groupthink: Groupthink, put simply, is intersubjective agreement made more militant and more controlling. Groupthink is a practice within collectives that determine the group's dynamics and ideologies, and often, allow little or no flexibility as to when someone is permitted to deviate from the group theme or consensus. This is the practice that most of us believe is characteristic of restrictive political systems or militant organizations that allow little or no freedom of individual expression or deviation from official dogma.

---

[63] Another part of the problem is the plethora of scientific papers that are being published. For example, astrophysics papers at ArXiv.org (Cornell University Library Archive) have several sub-categories that together result in more than 1000 new papers each month!

However, this is only partially true, groupthink is prominent in societies and organizations that are generally considered open and free, and might include just about any collective association, such as corporations, political parties, religious entities, as well as social or ethnic identifications.

Here is a short list, devoid of details, of some of the primary symptoms of groupthink:

- Unquestioned adherence to the group's ideology.

- Unquestioned belief in the morality of the group's acts and beliefs.

- Rejection of any information that might conflict with stated positions.

- Opponents are considered too naïve or ignorant to appreciate the group's policies and opinions.

- Illusion of unanimity within the group and the false notion that silence is consent.

- Direct and often coercive pressure on dissenters.

- Certain individuals take on the role of true-believers, enforcers, and protectors of the group.

Perhaps, after scanning this list, the reader might recognize that perhaps they too are a member of some groupthink association. Or possibly they might believe that this list belongs to *other people* and certainly not oneself! Of course, this list is not inclusive, and not all items listed are necessary for groupthink to exist. Only two or three items may be sufficient for the existence of groupthink.

There is also the question of how many members (or percentage of members) of a group must agree to a commitment such that it becomes a group

commitment? Other than the simple *it depends* answer, there is no simple or definitive answer. Group
hierarchy naturally plays a role as does how vocal are
the supporters of a commitment.

Within every group, committee, or social aggregate
there exists the group agenda that is public, the individual agendas of group members, and hidden
agendas. Additionally, when it comes to group dynamics, all members are never equally weighted: 1) There
are always groups leaders, implicitly or explicitly. 2)
There are always members who are stronger and/or
more vocal, and 3) there are those members along for
the ride.

It should also be noted that groupthink can apply
where there is no formal organization but only a lifestyle or other affectation that has followers who
indorse and adopt a particular lifestyle or sub-cultural
affectation. For example, those who follow a certain
fad or effect a certain mode of dress, overall appearance, and/or affected mode of speech and behavior.[64]

Social Epistemology and Scientific Truth

Without doubt, the word epistemology is, not a
word most of us, use on a daily basis, and many,
perhaps, are not exactly sure as to its meaning.
That is understandable, because even among
those who do use this word there are many

---

[64] Those interested should consult Janis, I. L. *Groupthink: psychological studies of policy decision and fiascos*. Boston, MA:
Houghton Mifflin, 1972.

different interpretations of its exact meaning. Some simply define epistemology as the study of knowledge and justified beliefs, and leave it at that, while others extend this definition to consider whether knowledge is an individual occupation or a collective occupation. Still others extend this definition to include the sources of knowledge along with consideration about the credence and veracity of these sources.

To be sure, what constitutes knowledge is not that easy to define, but it is clear that followers of *Social Epistemology* (hereinafter SE) want to change what we call *knowledge*. In any case, the term SE, being a relatively new notion, also has many interpretations.[65]

As the word *social* might imply, practitioners of SE tend to believe all human actions and interactions, including science, should be tempered with value-based considerations such as social agendas and, in some cases, personal agendas. Many advocates of SE state that science should include value-based considerations, which really means the use of purposeful exclusions or counterfactuals is acceptable and even desirable. This notion subordinates facts and truth to the whims of social and personal agendas, and this, in turn,

---

[65] Although the origin of the term social epistemology dates from the 1950s, it was not until Alvin Goldman and Steve Fuller became active in the 1980s that the term began to gain in popularity.

implies that it is legitimate to subordinate empirical truth

Some SE advocates even argue that scientific facts
are not discovered but manufactured. That is, science does not discover truths about the natural
world, science invents them. And, strange as this
may seem, this notion seems to be increasingly
accepted by many scientists today, who apparently   believe   that   epistemically   inferior
alternatives are allowable where goals other than
advancement of knowledge are present. Reverse
this, and it means that instead of advancement of
knowledge as a goal, it is legitimate for science to
have goals biased by personal and social agendas
that result in counterfactual statements or otherwise inaccurate information.[66]

It must be admitted that there are fragments of
truth in these suggestions. For example, science
does create conjectures about the workings of nature   as   a   means   of   explaining   natural
phenomena. However, these explanations, if
honest, are not arbitrarily conceived and, traditionally,   have   been   independent   of   social
agendas. Scientific conjectures and theories are

---

[66] Among other things, the word counterfactual, which is popular
among philosophers, can mean expressing a falsehood as if it
were the truth – usually intentional but could be un-intentional.
To this I might add that the online Oxford English Dictionary de-
fines counterfactual as; "*Relating to or expressing what has not
happened or is not the case.*"

created in an effort to explain natural phenomena in a manner understandable in ordinary language, and not to engage in deceptive sociallymotivates manipulations.

As an example of traditional scientific methods, when a physicist speaks of an *electron*, it is a reference to a set of phenomena, which, collectively, are attributed to *some-thing* called an electron. This *phenomenal electron*, far too small for direct observation, can only be *measured* indirectly by inference and extrapolation. That is, what we call an *electron* is the sum of a particular set of experimentally measured phenomena (data), which collectively has been given the name *electron*, and is representative of the behavior of this thing we call an electron. The explanations of science in common language necessarily entails human interpretation, and human interpretation entails assigning symbols or names as a means of individualizing associations and establishing familiarity.

Having said that, it should be noted that in the past care was taken to minimize any subjective content within a scientific conjecture or theory, but as more progressive social agendas gain in popularity, SE has become more influential and prevalent within the scientific community.

*   *   *

Another issue is that the use of social-political values is usually not explicitly stated, and the value modified empirical content is presented as completely valid and reliable empirical evidence.

Traditional epistemic values protected testimonial integrity, while the inclusion of social-political values subordinate empirically valid data (facts) and are often merely a form of *wishful-thinking*. It is part of the challenge of accepting the world the way it is versus the *wishful-thinking* that supports notions of how one thinks the world ought to be. That is, what *ought-to-be* shaded by social and political values that may or may not be viable or desirable such that the immediate relief of an *ought-to-be* change becomes problematic in the long-term.

For example, some sciences, such as astronomy, rely on assumptions based strictly on observational data, which is interpreted in a manner such that later they claim the data confirms the conjecture. All this is associated with the question of evidential adequacy, which varies by scientific discipline as well as individual opinion and, these days, is subject to manipulation to meet non-scientific agendas [The reader will recall the discussion ion the section *Scientific Theories of Today* – page 69].

The aim of science is to seek what is true about the world through formal means and methods, with results that are judged by being *truth-to-fact*. If science accepts values other than seeking truth, then science has diminished itself and is no longer free from social bias. This notion of truth-to-fact is an ancient tenet of science that was held by persons and societies long before there was science as we know it. That is, truth-to-fact was valued in those societies that allowed dissent and was tolerant of ideas that ran counter to the status quo. Truth is the paramount virtue of science and if corrupted, by weak or soft virtues then science is no longer science, it has become merely another form of propaganda [See *Side Comment on Virtues* following this section].

Advocates of SE, apparently believe that such modifications and censorship are justifiable, and that social values should play a role in essentially all human affairs, and not just science. As a result, personal, social, and hidden agendas have become well entrenched within society in general, and through a type of scientific groupthink that, discouraging heterodoxy, insists on a dogmatic adherence to certain agendas that include the increased inclusion of personal and social values into scientific research and theory acceptance.

* * *

There is also an interpretation of SE that focuses on the truthfulness or falsity of held opinions. For example, there is generally more credence given to testimony from someone known or believed to have expertise in the chosen topic than someone who does not. That is, known (or believed) competence adds credibility to a person's statements but such testimony should be considered valid only within their area of competence – otherwise it is just another person's opinion.[67] This use of

---

[67] In the field of Information Technology there is an old insider joke, that an expert is someone from out of town.

supposed competency is used in advertising of all types through the presentation of supposed expert testimony through the use of individuals, who are famous or have credibility in some field, to make statements about subjects totally unconnected with their supposed area of fame or expertise.[68]

In fact, advertisements, both political and commercial, are constantly making claims that appear superficially valid, but on closer examination are found to be fallacious. One of the most common errors they make, on purpose, are category mistakes, where there appears to be a connection, but none really exists.

In summary, the SE doctrine, essentially says that it is allowable, and even desirable, to ignore or modify physical facts to meet social agendas. Put another way, this means that empirical or factual knowledge modified or filtered by social agendas, or even the use of counterfactual content, is an acceptable practice. Put candidly, SE states that it is acceptable to withhold the truth and to lie.

Side Comment on Virtues: Virtues can be divided into two categories such as the *hard* virtues and the *soft* virtues. The hard virtues, which seem to have been more dominant in the past, include courage, duty,

---

[68] This reminded me of this famous quote by the humorist Will Rogers: "Everybody's ignorant, just on different subjects."

honor, honesty, and self-sacrifice., while, on the other hand, the soft virtues, which seem more dominant today, include pity, mercy, sensitivity (more inward than outward), and self-interest. There is nothing intrinsically wrong with either set of virtues unless they are taken to extreme. That is, it is not that the soft virtues are wrong or represent misplaced emotion, indeed, they are mostly positive and have moral value, if used properly, and the same can be said about the hard virtues.

Like so many other aspects of life both sets of virtues are usually mixed within an individual such that neither dominate, plus there are appropriate times when one set over the other will be more dominant. Given the relatively softer life of most who live in developed countries, I believe it is safe to say that today many, if not most, are shocked at the supposed lack of sensitivity and cruelty of our ancestors – who seemed to favor the hard virtues. However, it also seems apparent that they would be shocked by laxity, softness, and over self-sensitiveness of today's society.

Miscellany

Dynamic Accommodation

Throughout our lives, we are constantly acquiring new knowledge of the world. Even though we execute most of our daily rituals without much thought, in a subtle but very real sense, every time we perform, even a routine task, it is a new experience. Things have changed between each repetition of the experience. All sorts of things have occurred, and the world has moved on. For example, I visited the same coffee house this morning that I visited yesterday, but today's trip was different from yesterday's trip in a number of ways. The traffic was different, different people came into the coffee house, and so on.

Today, like all days, is similar and yet different from yesterday and every other day. The differences are only slightly noticed and the same can be said about the similarities. That is, in the course of our lives, similar-differences and different-similarities are generally overlooked or given little overt thought, if any.

This is because as we go about our daily lives, we dynamically accommodate these subtle differences, and unless we encounter a situation, which is traumatic or in some other way unusual, we tend to ignore subtle changes and in doing so, we dynamically accommodate these changes and receive emotional comfort through familiarity and the maintenance of normalcy.

Intersubjective Agreement

There is a class of agreements within societies that are a part of socially constructed reality and are based on implicitly recognized common agreements. These agreements can be as limited as an agreement between two persons or as broad as an agreement encompassing whole societies.

An intersubjective agreement is a collaboration and corroboration between individuals, which exists by extension within communities and societies in general. This type of agreement, in conjunction with the previously discussed dynamic accommodation, is not only essential for normal functioning, but is an essential element of societies and nations. That is, these agreements range from personal conventions to national conventions and some global conventions.

Being socially and culturally derived, these intersubjective agreements are usually non-binding such that they rely mostly on common understandings, social conventions, and are not governed by legal statutes. For example, at the level of societies, these types of agreements govern what is considered proper behavior or tacit agreements of confidence in certain aspects of society, such as confidence in the honesty of the election process, and the observance of locally accepted social practices. At the more personal level, married couples frequently rely on this type of agreement, although in this case, the agreement might be considered binding.

Moreover, these agreements often serve as a rein-
forcement or confirmation of believed reality and
provide a level of assurance that an individual's
personal reality is also that of the collective.

Groupthink

Groupthink, put simply, is intersubjective agreement made more militant and more controlling.
Groupthink is a practice within collectives that
determine the group's dynamics and ideologies,
and often, allow little or no flexibility as to when
someone is permitted to deviate from the group
theme or agendas. This is the practice that most
of us believe is characteristic of restrictive political systems or militant organizations that allow
little or no freedom of individual expression or
deviation from official dogma.

However, this is only partially true, groupthink
is prominent in societies and organizations that
are generally considered open and free, and
might include just about any collective association, such as corporations, political parties,
religious entities, as well as social or ethnic identifications.

Here is a short list, devoid of details, of some of
the primary symptoms of groupthink:

7) Unquestioned adherence to the group's ideology.
8) Unquestioned belief in the morality of the
   group's acts and beliefs.

9) Rejection of any information that might conflict with stated positions.
10) Opponents are considered too naïve or igno-
    rant to appreciate the group's policies and
    opinions.
11) Illusion of unanimity within the group and
    the false notion that silence is consent.
12) Direct and often coercive pressure on dis-
    senters.
13) Certain individuals take on the role of true-
    believers, enforcers, and protectors of the
    group.

Perhaps, after scanning this list, the reader might recognize that perhaps they too are a member of some groupthink association. Or possibly they might believe that this list belongs to *other people* and certainly not oneself! Of course, this list is not inclusive, and not all items listed are necessary for groupthink to exist. Only two or three items may be sufficient for the existence of groupthink.

There is also the question of how many members (or percentage of members) of a group must agree to a commitment such that it becomes a group commitment? Other than the simple *it depends* answer, there is no simple or definitive answer. Group hierarchy naturally plays a role as does how vocal are the supporters of a commitment.

Within every group, committee, or social aggregate there exists the group agenda that is public, the individual agendas of group members, and hidden agendas. Additionally, when it comes to group dynamics, all members are never equally weighted: 1) There are always groups leaders, implicitly or explicitly. 2) There are always members who are stronger and/or more vocal, and 3) there are those members along for the ride.

It should also be noted that groupthink can apply where there is no formal organization but only a lifestyle or other affectation that has followers who indorse and adopt a particular lifestyle or sub-cultural affectation. For example, those who follow a certain fad or effect a certain mode of dress, overall appearance, and/or affected mode of speech and behavior.[69]

---

[69] Those interested should consult Janis, I. L. *Groupthink: psychological studies of policy decision and fiascos*. Boston, MA: Houghton Mifflin, 1972.

Reconsidering Time and Space

"…one should not think simply in traditional forms of the explanation of nature, even if these forms are well formulated in historical philosophical systems but should be open-minded for new logical and empirical possibilities, not foreseen by these systems."[70]

As discussed in the above quote, opened mindedness is essential within the scientific community. History has shown that those who question the orthodox are usually the innovators. Giving up long held beliefs is never easy, but often necessary. We need to remember that stepping out of our conventional beliefs, like stepping out of the door, can lead to the unexpected. Just ask Alice…or Bilbo…

---

[70] Pauli 1994, 139.

Time Reconsidered

"It was shown already by Aristotle … that the psyche appears to be responsible for the 'measuring' and 'the counting' of (physical) motions and thus, for the coming into being and the genesis of chronos at the human level as a psychological phenomenon with ontological pretension."[71]

"How may it then be measured? And yet we measure times; but yet neither those which are not yet, nor those which no longer are, nor those which are not lengthened out by some pause, nor those which have no bounds. We measure neither times to come, nor past, nor present, nor passing; and yet we do measure times."[72]

As previously discussed, we measure common time by observing the change in position of an object, such as the *hands* of a clock. All objective measurements of time require physical motion, whether it is the daily passage of the sun or sand falling within a glass enclosure, the measurement of time has always been through some cyclical *physical* process, while our understanding of time is a *psychological* process.

In this section, we dig deeper into the concept of time to determine its true nature and to do this it will be necessary to let go of the notion of

---

[71] Attila Grandpierre, "The Dynamics of Time and Timelessness," *The Nature of Time: Geometry, Physics, and Perception.* NATO Science Series II. Mathematics, Physics, and Chemistry - Vol. 95, (Netherlands: Kluwer Academic Publishers, 2003).

[72] St. Augustine of Hippo, 129.

common time and open ourselves to novel ideas, and like all well entrenched and seemingly intuitive concepts, letting go of common time may not be easy.

Order

Before continuing with time, it will be useful to have a brief discussion about order. Again, like time, here is a common word used without much thought about what it really means. Although order has many definitions,[73] the sentential context usually provides enough clues to determine the intended meaning. Even though there are many types of order, in its simplest form, we tend to think that order has something to do with the arrangement of physical objects. For example, objects in a straight line are said to have order.

Another example is the sequential ordering of events, which, it seems, is an essential element of our understanding of common time. We build up our common notion of time from a sequence of events, which we perceive as having relational order. These events appear to occur in a continuous sequence that leads us to think of time as having a certain continuity. Sometimes, we call this sequential order the flow of time and,

---

[73] The online *Oxford English Dictionary* list about 50 usage definitions for the word *order*(2018).

sometimes, this flow is associated metaphorically with flowing water.

Our conceptual understanding of order is not intuitive but is one of the first things we learn. Starting in early childhood, parents and others teach order through the arrangement of things. First, just by lining things up and later as we develop language skills, our sense of ordering becomes more complex and terms like *alphabetical order* and *numeric order* become familiar.

When establishing an order, whether of events or objects, we learn to understand and use the relation-concepts of *before, after,* and *in between.* The first two are binary relations between two objects or events and the third is a tertiary relation between three objects or events. For example, if event A is *before* event B, and event C is *after* B, then B is *in between* A and C. These learned notions are so familiar that we rarely give them any thought and are examples of conceptual thought, which will be discussed later. These notions about order are well-known and learned at an early age, nevertheless, something is still missing.

We may know the sequence of events, but we really do not know anything definite about the intervals or gaps, which separate these three events (A, B, C) such as - are the intervals between these three events equal? We do not really know unless we have some means of measuring these intervals. Just as we might use a ruler, which is a standard linear unit of measurement, to measure an object's extension or the separation between objects, we need something equivalent to the ruler that will measure the intervals between events.

What Does a Clock Measure?

"The concept of a clock enfolds all succession in time. In the concept, the sixth hour is not earlier than the seventh or eighth, although the clock never strikes the hour save when the concept bideth."[74]

That may seem like a silly question since *everyone knows* what a clock does. It tells time. Of course, that is wrong; we only assume that it tells time. What it really does is something else, but before answering the above question, we first need to think about what a clock really is and how it functions.

Certainly, for everyday common usage, the definition of a clock as given, for example, by the online *Oxford English Dictionary* "as *a mechanical or electrical device for measuring time*" is completely adequate.[75] It only becomes inadequate when we delve deeper into the nature of clocks and time.

All clocks are devices that repeat some type of cyclical *motion*. Furthermore, we assume that this cyclical motion repeats at equal intervals. There

---

[74] Nicholas of Cusa (1401-1464), *Vision of God*, 1453, 24.

[75] To be sure sundials have no mechanical moving parts, it this case, it is the earth's movement that provides the measure of *time*.

is just no other way of measuring event intervals except with predictable repetitive motion associated with an object that is external to the object of measurement. That is, if I want to measure a series of events, I, necessarily, have to use a measuring device that is independent of the events being measured.

Clocks, in a monotonous manner, simply repeat mechanical processes with what we faithfully believe are equal intervals. These repetitive processes require an interpretative interface through which the repetitive motion acquires meaning. As Nicholas of Cusa wisely observed six hundred years ago, six o'clock is not a characteristic of a clock; but has only a symbolically interpreted meaning that we give to a particular configuration of the clock-human interface.

The human interface of a clock is only a visual mechanism of successive positions and is indifferent to the hour. Although we commonly assume that a clock is *telling time*, what clocks really do is provide a standard measure against which we determine, through interpretation of the clock-human interface, the intervals between events.

To be sure, the history of clocks is rich and varied, and today we encounter clocks in many configurations. Nevertheless, all clocks still have essentially the same purpose, and they all rely on processes that meet our expectation of reliably repeating a cyclical motion with equal intervals. That is, they rely on the repetition of a particular motion with intervals that are assumed to be equal within some acceptable degree of accuracy. This is what all clocks do, whether it is sand falling through a glass tube or a modern sophisticated *atomic clock*.[76]

Does Time Flow?

In the world of everyday experience, time, whether continuous or discrete, certainly seems to flow in a continuous manner and is sometimes compared analogously to the flow of water in a stream. That is, water in a stream appears to flow in an unbroken manner, but as we are aware, the water in a stream only seems continuous. Actuality, the stream is an aggregate of an extremely large number of water molecules and therefore, at the micro-level, this flow is discrete and not continuously connected.[77]

Direction of Time

The idea of the flow of time is still not complete. The flow needs a direction. At this point, you might be tempted to say that this is obvious –

---

[76] Even electronic clocks use *cyclical or oscillating motion* as the measure of interval.

[77] An 8-ounce (224 grams) glass of water has about 7.5 X 10^{24} molecules of water. (10^{24} = 1,000,000,000,000,000,000,000,000). To put this into perspective, there are more molecules of water in an 8-ounce glass than there are glasses of water in all the oceans!

time flows from the past to the future. Well, for all practical purposes, this notion of the flow of time seems correct, but as with other common usage, this statement needs closer examination.

If I move the cup on my desk from one position to another, I can always move it back to the original position. In doing so, did I reverse time, or did I simply move an object without any influence on time? About all we can say about the two moves is that we only changed the position of the cup. We did not change its appearance or condition. That is, these moves left the cup itself unchanged and after the second move, it was as if the cup never moved. Both cup moves are examples of *reversible* processes and will not help us determine the direction of time.

Instead of just moving the cup, what if I drop the cup to the floor and the cup breaks. We can all agree that the cup is not likely to jump back onto the table and become a whole cup again. This is a very different situation from just moving the cup from one position to another and back again. The breaking of the cup is an example of an *irreversible* process. Since the cup went from an unbroken state to a broken state. It may well be that irreversible processes might be useful in determining the direction of time.[78]

---

[78] The direction of time is sometimes called the arrow of time. For a good discussion of various theories on the arrow of time I recommend the *Stanford Encyclopedia of Philosophy*, http://plato.stanford.edu/.

Past, Present, Future

"Predictions are difficult, especially about the future."[79]

In everyday experience, we often encounter the terms past, present (or now), and future. And once again, just as with the common words previously discussed, we use these words with little thought or understanding of what they might really mean.

The future: Of these three words, future is the least problematic. We anticipate the future based on our experience of past events, our expectation of future events, and our wishes (anticipated desires). Future events are undecided and have probabilities of occurrence running from extremely high to extremely low. The former being events such as the certainty that the sun will reappear in the morning and on schedule, to the latter which includes low probability events such as winning a lottery.

In the middle, somewhere are undecided and uncertain events where with reasonable degrees of probability we anticipate and expect a resolution. This includes events such as athletic competition, where an outcome is certain, but the exact outcome is speculative. Here we have an example of a subtle and important distinction regarding the status of certain future events. An event that is

---

[79] Quote attributed to Niels Bohr, a Danish physicist and one of the founders of quantum theory.

determinate – it will be settled, but not pre-determinable – results not known in advance.

The future, then, is this realm of possible events, which we look forward to with anticipation, anxiety, and sometime fear. The future is unrealized potential.

The past: Events of the past are fixed in the sense that they cannot be altered, and, for most of us, events of the recent past are not problematic in the sense that we are capable of recalling actions and events of the recent past with an acceptable degree of accuracy. For example, finding an object today that was placed somewhere yesterday is usually quite easy. The past, when it concerns recent events, is usually very neat and tidy.

On the other hand, the reliability of memory for events in the more distant past can be problematic. It is unfortunate and often inconvenient truth that memory – like the oral traditions of ancient times – is not very reliable. This is because, either consciously or subconsciously, we regularly alter our memory of past events.[80] For example, we often modify unpleasant and emotional memories to make them more acceptable

---

[80] I have purposely used the word subconscious because the word unconscious is often associated with not being awake or lack of awareness. Unlike Freud, I have no problem understanding that subconscious does not refer to spatial sub-structure but to a hidden part of the Mind.

and tolerable. Of course, sometimes, we just forget.

Mostly, we base our future actions on remembered experiences based on often modified or prejudiced recollections. That is, our prejudices - for indeed we all have them - unavoidably influence our actions. The past is fixed, but recollections of the past are not and, besides, no two persons share the same past.

The present: Briefly, the present is the fleeting middle ground that is wedged between the past and the future, which on recognition as the present has already become the past. Since the *present* is where we seem to spend our lives, it is somewhat more complex and elusive than future or past, so we will need to dig deeper to see what is really going on.

What now?

The truth of the matter is that everything we experience has already happened. There always exists a delay in the reception of sensory sensations and an additional delay in the comprehension of this data. For example, whether the perceived event is a visual image, a sound, or a combination of both, these events have a short but real delay due to the finite speed of light and the much slower speed of sound.

Additionally, there is the delay of comprehension that must follow the reception of the sense data.

In this way, the *now* has already slipped into the recent past before we are completely aware of it. Of course, this happens very quickly so that the delay is subtle to the point of being unnoticed and most events give the impression of being immediate. This essentially means that there is no *now* and that all references to the present or now are always somewhat vague. Such references are only appropriate for informal or common discourse.

However, the notion of the present is deeply ingrained in the human psyche, and all efforts to change our ideas about the present are, at best, difficult or possibly even futile. Perhaps, the best we can do instead of thinking in terms of some instantaneous now or present, is to modify our idea of the present and adopt an extended version of present. That is, to smear it out – like passing a finger over ink not yet dried.

Duration

"Duration can be all nature present as the immediate fact posited by sense-awareness."[81]

"I will use the term 'moment ' to mean 'all nature at an instant.' A moment, in the sense in which the term is here used, has no temporal extension, and is in this

---

[81] Alfred North Whitehead, *The Concept of Nature*. (Cambridge: Cambridge University Press, 1920), 24.

respect to be contrasted with a duration, which has such extension. What is directly yielded to our knowledge by sense-awareness is a duration."[82]

"Duration is the continuous progress of the past which gnaws into the future …" [83]

In the course of our daily lives, it seems apparent and intuitive that events flow continuously. That is, it appears that each moment connects to the next without a gap so that the movement of an object appears continuous. Thus, we perceptually conceive that the passage of events through time and space is unbroken. However, this apparently continuous flow is an illusion. What we really experience is a sequence of images or durations that we mentally interpreted as being a continuous flow.

That is, instead of a continuous flow, we experience a sequence of discrete durations where the current duration is sandwiched in or delineated by the beginning of the next duration and the end of the prior duration. This ensures sequential continuity between previous durations and future durations.

Perceptually, the flow of these durations smears out, or run together, with no definite beginning or end. These sequences of smeared out durations comprise what I will call the specious present and through this continual sequence of

[82] Ibid, 57.

[83] Bergson 1911, 5.

durations we experience both the flow and direction of common time.

Specious Present

In 1892, the eminent American psychologist and philosopher, William James (1842-1910) was able, using some existing and some novel ideas, to redefine the present in a form that he called the *specious present*.[84] He defined the specious present as, "a sort of saddle-back of time with a certain length of its own, on which we sit perched, and from which we look in two directions into time." Although some have questioned the use of the word specious, it seems appropriate since specious is defined as "superficially plausible, but actually wrong or fallacious."

That is, the specious present is the *window* through which we *observe*, both forward and backward, the sequence and succession of durations. This is similar to the way we interpret musical notes, not as individual tones, but as a flowing sequence of notes, which seem continuous rather than individual. Additionally, in the case of familiar music, the anticipation of future notes also contributes to the apparent continuous flow. Essentially, we trick ourselves into thinking of music as a continuous flow.

---

[84] William James, *Principles of Psychology the Briefer Course - 1892* (Cambridge MA: Harvard University Press, 1923), 280.

The same holds for any sequence of events, which occur at a rapid enough pace to seem continuous, just as the pace of each frame of a video appears in our consciousness as an unbroken flow. Both the apparent flow of music and the continuous flow of a video are dependent on a certain amount of perceptual persistence and on our minimum threshold of perception.

The specious present builds on this flow of durations, which results in perceptual continuity called the passage of time, but more correctly, should be called the passage of nature. The passage of nature can be as short and fleeting as a flash of lightning, or more drawn out such as the growth of a garden. The passage of nature is our awareness of the world moving through our consciousness. It is the flow of our lives.

Tensed or Tenseless Time?

In 1908, J. E. McTaggart (English philosopher, 1866-1925) wrote an article about time for the quarterly journal *Mind*,[85] in which he defined two categories for time, one tensed and one tenseless. He called the tensed category the A-Series, which included the concepts of *past, present (now)*, and *future*. He called the tenseless category the B-

---

[85] John Ellis McTaggart, "The Unreality of Time," (1908)
http://www.ditext.com/mctaggart/time.html.

Series, which included the relational concepts of *earlier than*, *later than*, and *simultaneous with*.

The A-Series is essentially the same as what here we call common time, which might also be called chronological time or clock time. Additionally, the A-series has a *quasi-absolute* element because there was only one occurrence of 10:15AM Pacific Standard Time on April 21, 2019 in San Francisco – quasi-absolute because our calendaring and time-keeping processes are artifacts of convenience not natural things or events.

The only way we experienced the A-series (tensed time) is through notions about before-after sequences of events. We remember past events and we anticipate future events, while through the specious present these events flow from the unrealized future into the remembered past and only within a fleeting duration are these events perceived. There is nothing new or strange about the A-series. However, the B-Series is different, it has no absolute reference point, and is independent of when and where.[86]

The B-Series does not include the concepts of past, future, or present and makes no distinction about when or where something happened. Essentially the B-series measures intervals – a beginning and an ending – and determines order and sequence. For example, a reading of a clock

---

[86] This paper also discussed something McTaggart called the C-series, which he called a series of permanent order and is non-temporal.

at the start of a scientific experiment is compared to another reading made at the end of the experiment. These *readings* are taken from some sort of clock, which are then inserted into the experimental calculations and interpreted as an *interval of time* – although the actual measurement was a *clock interval* (the difference between two clock readings). The B-series is like the timing of an American football game – where small timed intervals accumulate until one hour is reached and the game ends.

Of course, today's experimental intervals are most likely measured by a computer's internal clock (the internal cyclical mechanism that all computers use for synchronization). The B-series is useful in creating a sense of order through such relations as *earlier than*, *later than*, and *simultaneous with*. These relations specify order and sequencing for events as they occur in our everyday situations.

Time Slips Away...

For most of us, time is an unseen and unknown entity, which keeps an indexical pace with events. However, time is actually a convention of convenience that we use in place of change and motion. We can consider money as being somewhat analogous to time. We use money because it is convenient and much simpler than trading in hard goods or services. Today currencies are not backed by precious metals and have value only through common agreement. Currency is simply an indexical measure used to relativize and equate values, and, besides, today most *currency* only exists as computer data.

The notions of time abstractly symbolize the physical aspects that we associate with change, motion, and concepts such as interval. Time is not the cause of any of these; it is not the cause of anything, with the possible exception of anxiety and anticipation. Put another way, time is defined through change and motion, not the other way around. Only through change, in the broadest sense, does time have meaning.

Starting in the late 18th century, and continuing throughout the 19th century, social and economic changes came fast, and as countries industrialized, they rapidly increased their dependence on the clock. As a result, the clock became a dominant factor in how we managed our lives. Under the influence of clocks, people began breaking the day up into smaller chunks.[87]

No longer dividing the day into morning, noon, and afternoon, now the common measure of the day is by hour and minute. Instead of meeting someone in the afternoon, it was meeting someone at 3:15PM. All this reliance on clocks changed the way we think about time.

---

[87] It should be noted that ship navigation was a strong motivator in the development of reliable clocks. The navigators needed these smaller chunks of time keeping, it was the only way navigators could accurately determine the ship's longitude.

What had been a loose concept based on natural events, became an indispensable concept driven by the needs of modern society. Consequently, our notion of time developed into a dependence on the clock that is both pervasive and deeply ingrained. As previously mentioned (*Common Time* – page 3), clocks provide a mechanism for dependable repetitive motion and we feel intuitively confidant that these devices keep a steady and invariant pace of measurement. Is this confidence justified?

When the special theory of relativity (previously mentioned on page 57) demoted time, clocks lost some of their credibility. For example, assume that we have two synchronized clocks setting next to each other, if one of these clocks suddenly accelerates away from the other, the accelerated clock will run more slowly than the one left behind and they become unsynchronized. This is sometimes called time dilation, which is the observed effect of acceleration on mechanical devices that are used to measure – intervals that we call - *time.* Among other things.

For example, consider for a moment the Global Positioning System (GPS) on which so many depend. This remarkable system would not be possible without the knowledge of both the special and general theories of relativity. Indeed, without compensating for these theories, the GPS satellite system would quickly be out of synchronization and become inoperative.

As mentioned previously, irreversible processes, and our notions about the flow of time present what appears to be a uniform (future-directed) progression of time. Although useful in ordering our lives, this progression, that we call time, is just a mental notion without physical reality. Relativity theory stripped time of its ontological status and demoted time to a mere epiphenomenon. As stated in the above quote by Grandpierre, time is "*a psychological phenomenon with ontological pretensions.*"

This is the time of everyday discourse such as when someone asks, "*hat time is it?*" or statements such as "*I do not have time for that.*" Such references to time are references to a psychological concept that we call time. On the other hand, there are technical references to time that often assume, or give the appearance of assuming, that time is a physical thing. That is, the notion that time has some absolute existence such as the air we breathe. This notion is usually accompanied by the additional notion that time is also a continuum of some kind such that there are no gaps between any two time-like measurements.

Put another way, time is a state of mind that we have learned to intuitively believe is something physical and associate it with our *outer intuition* (sense perception) such as the separation of events. The *inner intuition* (mental notion) of time has no geometric image other than that of a line. The physical sensations of events and motion, which define change, together with our mental notion of a linear sequence convince us that time flows in a definite direction.

It also seems that there has been a long-held belief that change is somehow interconnected or joined with time. That is, many believe that time is the cause of change. Change has many causes, but time is not one of them. This misconception of associating time with change, as well as other misconceptions about time is the result of believing time is a physical entity and part of *concrete* or veridical reality, and even though special relativity successfully refuted the notion of absolute time, the notion of time as some-thing physical appears to be as pervasive as ever – even among scientists who should know better.

Considering time as a requirement or cause of change casts time as some kind of substance with causal capabilities. That is, considering time as possessing some kind of concrescence implies that time is a physical force or causal agent of some kind, even though no physical evidence has ever been found – or will be found.

Change occurs through motion. That is, motion in the broadest sense, which includes everything from the movement of the pen when writing this sentence, to the movement of massive astronomical objects, to chemical reactions such as the rusting of iron, and the radioactive deconstruction of matter at the atomic level.

We know the physical (mechanical) variability of measured time through the theory of relativity and we know the psychological (mental) variability of the flow of time through personal experiences. The variability of time (or more correctly, the variability of our sense of the passage of nature) is, of course, a well-known experience.

All of us have experienced the fleeting sense of time during pleasant occasions as well as the painfully slow plodding of experienced time during stressful occasions. Of course, the clock on the wall is not bothered by either type of occasion, and its relentless and consistent motion is our only source of an unvarying flow of elapsed *time*. It is through the progression of individual events blending into a sequence of durations, the specious present, and eventually into our sense of the passage of nature that creates our personal notion of time, which is never consistent and steady [Recall the section *Natural Time* - page 4].

*   *   *

Given the many theories and numerous volumes that have been produced dealing with time, it may seem abrupt and presumptuous to dismiss time completely in only a few paragraphs. However, proper reflection on the nature of time makes this the only choice. Philosophers and many others have long debated the reality of time but usually they start with the notion that time is a some-thing and then try to prove or justify this choice -the approach herein has been to investigate first then make a choice.

To add to the confusion, even after more than one hundred years of special relativity, many authors of technical papers, still have not let go of the notion of time as being something physical. Sometimes the same author will acknowledge that time is a phantom, then later use phrases that imply that time is a physical thing, such as *time itself, time beginning or ending, static space-time,* and so on. Other terms used for time include *operational* time, *proper* time, and *physical* time. Additionally, most often, these references to some-thing called time are used without definition such that the authors apparently assume that the reader either knows the definition or can contextually guess the intended meaning.

Understanding the reality of time may not influence our daily activities but when it comes to contemplating fundamental questions about life and the universe the true nature of time must be considered. That is, considering time in the usual way is a barrier to fundamental thinking about reality and the universe. Paradigm changes are essential for progressing in our understanding of the natural world and progress only occurs when certain *facts* or *beliefs* are put aside and replaced by novel ideas and concepts.

Side Comment on Scientific Time: Time, as used in science, is a mathematical convenience. For example, time in classical physics is completely reversible. That is, replacement of the variable t [=time] with –t is always permissible, but less often noted is that this t can also be replace by a function such as $t = f(s)$, where $f(s)$ is an arbitrary function or, perhaps, some mechanical process. Here is a quote by one of my former math professors; "*A parametric representation of a curve is not unique. Indeed, there are infinitely many representations, as shown by the fact that it may be replaced by an arbitrary single-valued function of a new parameter s, say t = f(s)...*"[88] This is allowable because, when the variable t is used for time in an equation it represents an interval measurement. That is, a B-series measurement of the difference between successive readings of an instrument called a clock.

Instead of stating that a certain elapsed time has occurred during the experiment, a function defining the clock's mechanism and mechanical motion could have been used. That is, the measurement of interval is really a count of the number of clock cycles that occurred during the interval. Time could also be replaced by a function representing a linear segment. Since these and other options are cumbersome and less convenient, the variable t is used.

---

[88] C. E. Springer. *Tensor and Vector Analysis with Applications to Differential Geometry*, Dover, NY, 2012,13.

Space Reconsidered

"There is no absolute space, and <u>we only conceive of relative motion</u> [emphasis added]; and yet in most cases mechanical facts are enunciated as if there is an absolute space to which they can be referred."[89]

"There is no entity 'physical space;' there is only the abstract space chosen by the physicist as a structure in which to plot phenomena; and some choices give simpler theorems than others (thus making the laws of nature look simpler)." [90]

"Space, like time, would appear to be an abstraction from events. According to my own theory it only differentiates itself from time at a somewhat developed stage of the abstractive process. The more usual way of expressing the relational theory of space <u>would be to consider space as an abstraction from the relations between material objects</u> [emphasis added]." [91]

Maybe these three quotes give away the punchline for the theme of this section but, just in case, let us proceed as if they did not and see if these statements are justified.

Over the centuries, space was considered an absolute entity. That is, space was thought to exist as some kind of physical object, a sort of container within which resided the earth and the other

---

[89]Henri Poincaré, *Science and Hypothesis* (New York: Dover Publications, 1952, English Translation 1905), 51.

[90] Milne 1951, 27.

[91] Whitehead 1920, 37.

heavenly bodies. Of course, if the universe is in a box then what contains this box and so on!

As mentioned previously, the geometry of this space was the *flat* geometry of Euclid and it dominated thinking about space for more than two thousand years. This belief in Euclidean space has been so deeply ingrained that it has only been within the last two hundred years that new and radical notions about space emerged that questioned the validity of Euclidean space.

Early in the 19th century, a couple of mathematicians took exception to Euclid's notions of flat space and to his parallel postulate (axiom) in particular.[92] These two mathematicians, Janos Bolyai (Hungarian, 1802-1860) and Nikolai Ivanovich Lobachevsky (Russian, 1792-1856), each, independently, began to develop what we now call non-Euclidean geometry, which is any geometry that deviates from the original Euclidean postulates. What they developed is today called *hyperbolic* geometry and is one of the two most common examples of non-Euclidean geometries. The other common non-Euclidean geometry was developed about the same time by the mathematician Bernhard Riemann (German, 1826-1866) and is called *elliptic* geometry.

---

[92] This postulate essentially states that given two straight lines, both perpendicular to a third straight line, are parallel in the sense that these two lines will never intersect no matter how long they extend. There are various versions of this postulate but this one works.

Here is a straightforward way to visualize these three types of geometry; think of the Euclidean as the flat surface of a basketball court, elliptic as the surface of a soccer ball (or American football) and hyperbolic as the shape of a Pringle potato chip.

In any case, the invention of non-Euclidean geometry, in the first half of the 19th century, received little attention through the rest of that century. Almost everyone considered this unusual geometry merely a mathematical curiosity with no connection to the real world. It seems that the 19th century, when it came to space, was more interested in things like measuring the speed of light and filling space with some mysterious substance called *ether*.[93]

Universal Expansion?

"…if the foundational issues have not been completely clarified from the outset, and if a rigid set of problems has been accepted without enough discussion about their relevance, one's lasting inability to solve some of these problems can be perceived as a symptom of permanent "crisis" … Yet this so-called "crisis" of the theory may well boil down to <u>a crisis of the too quickly accepted formulation of its problems, or to a crisis of</u>

---

[93] This is the so-called *luminiferous aether* and should not be confused with the ether once used as a general anesthetic.

<u>the prejudice associated to it in the name of immediate efficiency</u> [Emphasis added]."[94]

"The observations may be fitted into either of two quite different types of universes. If redshifts are velocity shifts, the model is closed, small, and dense. It is rapidly expanding, … On the other hand, if redshifts are <u>not</u> primarily due to velocity shifts, the observable region loses much of its significance. <u>The velocity-distance relation is linear; the distribution of nebulae is uniform; there is no evidence of expansion, no trace of curvature, no restriction of the time scale.</u>"[Emphasis added][95]

"The assumption that redshifts measure the rate of expansion of the universe is <u>a long extrapolation</u> from the familiar small-scale Doppler effects. Clearly, the data should be analyzed with the aid of <u>as few assumptions as possible</u>. The observations must be <u>corrected for energy effects</u>, regardless of the origin of redshifts. The interpretation of the corrected data is then direct, economical, and very simple."[Emphasis added][96]

The first quote is a warning that premature acceptance and hasty judgments can lead to fundamental issues, which because of these incorrect assumptions and acceptances, can lead to

---

[94] Michel Bitbol, "Reflective Metaphysics: Understanding Quantum theory from a Kantian Standpoint," (2010)
http://logica.ugent.be/philosophica/fulltexts/83-3.pdf

[95] Edwin Hubble, "Effects of Red Shifts on the Distribution of Nebula," *American Astronomical Society* – Provided by NASA Astrophysics Data Systems. (1936) (http://adsabs.harvard.edu/cgi-bin/basic_connect?qsearch=edwin+hubble&version=1), 552.

[96] Ibid, 553.

permanent creative crisis and underdetermined theories such as those to be discussed in this section.

The second and third quotes are by the American astronomer Edwin Hubble (1889-1953), who was an early 20th century pioneer of extragalactic astronomy. The first Hubble quote was chosen because it mentions the two most common interpretations of the redshift - velocity or distance - and it shows that, early on (1936), there were serious doubts about the meaning and proper interpretation of the so-called cosmic redshift.

In his second quote, from the same paper, it seems that Hubble's skepticism is even more apparent. He expresses concern about *a long extrapolation*, *as few assumptions as possible*, and that the data needs *corrections for energy effects*. These seem like the comments of someone expressing both concern and caution when interpreting the redshift phenomenon.

*   *   *

Side Comment on Redshift: Redshift is by definition an increase in wavelength and since photon energy is proportional to the wavelength, the longer the wavelength means less energy. That is, redshift indicates loss of energy in transit. These shifts in frequency are determined through spectrographic analysis and the resulting spectrographic images rely on the fact that each chemical element (oxygen, nitrogen, etc.) has unique electromagnetic spectral absorption pattern, and astronomers use these patterns to uniquely identify stellar and galactic chemical composition. If these patterns are out of position, then a shift in energy has occurred. Put simply, the light from distant galaxies are compared with light from the normal spectral patterns of light sources here on earth to see if the distant light has been *shifted*.

*   *   *

Anyway, before we continue, we need to back up a little. In 1912, the American astronomer Vesto Slipher (1875-1969),[97] while studying spectrographic images from various nebulas (now called galaxies), was the first to notice the phenomenon that became known as the cosmic redshift. This discovery drew little attention over the next few years, and it was not until better observing instruments became available that the redshift began to receive significant attention. In fact, it was the above-mentioned Hubble, expanding the work of Slipher, who made redshift a hot item.

Fast forward to the 1920s, Hubble used the new 100-inch reflector telescope at Mount Wilson in California to study the redshift phenomenon and by the late 20s had developed what, unfortunately, became known as Hubble's Law. This *law*, as originally formulated, related redshift with distance, but, later, to support the claim of an expanding universe it was re-interpreted as a

---

[97] In 1912, Vesto Slipher was an astronomer at the Lowell Observatory in Flagstaff, Arizona.

relation between redshift and velocity.[98] Although, as noted above, it seems that Hubble himself was somewhat uncertain about this idea. In any case, the reinterpreted redshift has played a significant role in the development of what is now called the Big Bang (hereinafter *BB*).

* * *

Side Comment on Expansion: The interpretation used by the expansion conjecture is that when the lines are shifted toward the *red* (lower frequency and lower energy), the object is assumed as moving away, and when shifted toward the *blue* (higher frequency and higher energy) the object is assumed as moving closer. That is, this conjecture is analogous to the well-known doppler effect that we experience with sound waves.

* * *

The expansion and *BB* conjectures started in 1927, when Georges Lemaître (Belgian, Catholic Priest and physicist, 1894-1966), used the redshift as a velocity measure to suggest that the universe was expanding. Later, in 1931, Lemaître developed this expansion conjecture into the idea that the universe started with an explosion of something he called a *primitive atom*. This was the beginning of universal expansion and the *BB conjecture*, although it was not until 1949 that the term *BB* was first used. English astronomer Fred Hoyle coined

---

[98] This reassociation committed the logical fallacy of *concurrence does not imply causation* or invalid *cause and effect* error.

the term *Big Bang* during a 1949 BBC radio broadcast - it was meant to be derogatory.

Over the years, the *BB* and expanding universe conjectures (hereinafter *BB/E*) gained in popularity to the point of being considered by many, if not most, astronomers and astrophysicists as a confirmed reality. On the other hand, there is still a large and possibly growing group of skeptics who are not sure all is right with the concept. This is because, over the years, the *BB/E* conjecture kept running into trouble. However, instead of looking in a new direction, it has always been decided that all the *BB/E* needed was another patch.

For example, in the 1970s, there was a realization that the current *BB/E* narrative did not *grow* the universe in quite the right manner. Specifically, it was realized that very soon after the BB that fine-tuning was required for the universe to appear the way it actually does. However, since some possible choices for fine-tuning might have certain unwanted philosophical implications, the *right kind* of tuning had to be *discovered*.

Then along came Alan Guth[99] who, in 1980, suggested that almost immediately after the *BB*, the universe began to expand much *faster than the speed of light,* that is remarkable enough but even more remarkably, expansion lasted only about *one ten thousandth of a second*, then the *inflation* (as

---

[99] In 1980, Alan Guth was an astrophysicist at MIT.

it came to be known) suddenly stopped as it reached just the right degree of expansion!

How convenient, a perfect fit! This convenient *discovery* (made-up story?) saved the day for the *BB/E* enthusiasts and Guth's idea was quickly accepted as gospel. This was so even though Guth's proposal strongly violated the special theory of relativity provision that the speed of light could not be exceeded by any particle that has rest mass greater than zero (let alone the whole universe at that time!)![100] So, the BB/E notion was patched up again and with the addition of cosmic inflation it grew in scope to become the BB/EI notion.

However, the truth is, that the introduction of inflation was hand fixing and fine-tuning to the extreme, but somehow it managed to become widely accepted as an integral part of modern cosmology. Moreover, the inflation notion also included a generating elementary particle called an *inflaton*, which, not surprisingly, is yet to be discovered. It is easy to picture this ongoing search for the mythical *inflaton* particle with a bunch of latter-day *Argonauts* scurrying about in all directions searching for the *golden fleece* of inflation.

Perhaps the biggest problem with inflation is the dependence on the BB conjecture, which is itself

---

[100] In special relativity theory, the rest mass is essentially equivalent to what we normally call mass for an object at rest (not moving relative to its surroundings – such as a coffee cup sitting on a table).

an unproven notion and, as of today, there are *no confirming observations of expansion* and the privileged BB/EI is allowed *ad hoc* corrections that other theories are denied. This is only one example of the problems encountered and the fixes that, over the years, have been applied to *BB/EI*.

As another example of questionable conjectures that are claimed to support the BB/EI, there is the much-touted cosmic microwave background radiation that is claimed to be *proof* of the big bang even though it is possibly just *intra-galactic noise* and not *extra-galactic* as claimed.

The external world is presented to us much like a shadowy reflection of reality, which presents only a specific perspective. This is not to say that we live in the shadowy world of a *Platonic cave*, but that our limited ability to perceive reality constrains our knowledge of the actual world. Just as we only see the surface of a body of water, we know that there is a vast reality hidden below. Unable to see below, we make assertions about this hidden reality based on interpretive extrapolations that go far beyond the given. Together the BB/EI, and other conjectures - such as dark matter and dark energy[101] make up what is called the *Standard Model of Cosmology* (hereinafter SMoC),

---

[101] Dark matter and dark energy are hypothetical stuff that some say may represent as much as 85% of the total matter of the universe. Both were added into the SMoC – by hand - to patch it up. Neither have been found after many years of investigation.

which appears much like a leaky boat that has been patched too many times.

For example, physicist Pavel Kroupa (Czech-Australian astrophysicist , 1963 -) wrote a paper a few years ago that clearly illustrated the problems with the SMoC.[102] In this paper, he outlined 22 issues with the cosmology model that had arisen over the 32 years between 1980 and 2012. In this paper, Kroupa clearly identifies these many issues with the SMoC, which are either ignored or, as is sometimes the case, altered through spinning to look like a discovery. Examples of the latter are dark matter, dark energy, and inflation, which were all put in by hand to solve problems, but their *discoveries* (patches) were touted as worthy scientific achievements. Here is a quote from his summary: "*Put another way: that the currently popular SMoC needs to be based to more than 96 per cent on unknown physics is nothing more than an expression of our present-day ignorance of how mass, space and time unify, i.e. of cosmological physics.*"

Even with all these fixes, other factors seem not to have been completely considered, including; density of photons, loss of photon energy (redshift = tired light?), and the unknown impact of a

---

[102] Pavel Kroupa (2012), "The Dark Matter Crisis: Falsification of the Current Standard Model of Cosmology."
https://arxiv.org/abs/1204.2546v2

multi-billion year journey on light propaga-tion.[103]

The SMoC is composed of too many assumptions and too many unverifiable conjectures such that skepticism about the veracity of modern cosmology seems unavoidable. Modern cosmologists might look with amusement at the epicycles and geo-centric cosmological conjectures of the past, but are today's conjectures any better? It appears that much of current cosmology is more like fanciful or wishful thinking than empirical science.

The historical track record for astronomy and cosmology over the last 100 years is not very convincing. It is exceedingly naïve and arrogant to claim that in a very short time; we have – finally - been able to *get it right* and that the current SMoC is a *true and accurate description* of the origin and development of the universe.

We need only consider the history of cosmology to realize that, although there is a lot of data being collected and a lot of analysis, and even though cosmologists have been wrong over and over again, all this observation and analysis is followed by new notions that are then claimed as true and accurate. As someone once said; "*Cosmologists and astronomers are often in error, but never in doubt.*" True believers in SMoC seem to live in a fantasy world where they apparently

---

[103] Those interested in additional comments should refer to: Hartnett, J. G. (2011): "Is the Universe Really Expanding?" https://arxiv.org/abs/1107.2485

really believe their tales of exactly what happened and when it happened, but without a clue as to *why* it happened.

Beyond these doubts about the veracity of current cosmological conjectures are doubts about whether we are even capable of making sound cosmological conjectures. This doubt has been around for more than a hundred years and these doubts were voiced by some of the abler thinkers of their times.

For example, the famous Austrian physicist and philosopher Ernst Mach (Austrian physicist and philosopher, 1838-1916) held the view that the universe was not an object to which we could apply meaningful statements. Another example is that of the Pierre Duhem (French physicist and philosopher, 1861-1916) who believed that physical theories about the universe could never be confirmed or negated.[104] Examples that are more recent include British Astronomer M. J. Disney,[105] South African cosmologist George Ellis,[106] and the above mentioned Pavel Kroupa. This list

---

[104] The Mach and Duhem examples are taken from Helge Kragh, "The Controversial Universe: A Historical Perspective on the Scientific Status of Cosmology", https://d-nb.info/101824364X/34.

[105] M. J. Disney (2000), "The Case Against Cosmology", arXiv: astro-ph/0009020v1.

[106] George F. R. Ellis (2006). "Issues in the Philosophy of Cosmology. Section 2.8.1, Misconception 1," arXiv: astro-ph/0602280v2.

could go on, but this is enough to make the point that even today not all scientists are *true believers*.

As discussed above, astrophysicists have been putting patches on the *BB/EI* conjecture for many years and they seem determined to save this conjecture at any cost. Whether this is because they believe the conjectural basis is fundamentally sound or there are other reasons is difficult to say. However, possibly (probably?) these extreme efforts to save the *BB/EI* conjectures are due to the influence of hidden agendas.

There seems little doubt that the *BB/EI* and associated conjectures are, at least to some degree, ideologically motivated and this motivation is inclined toward reductionism and scientific realism. Despite all these attempts to rescue the conjectures using *ad hoc* patching, the BB/EI conjecture still fails to have predictability or any predictions that can be tested.

It should be noted, that in discussions about the expansion of the universe, it is often stated that the expansion is due to the expansion of space itself, not due to relative motion of spatial objects. This obviously assumes that space is something that can be expanded. That is, a physical entity of some kind. This is also sometimes given as the reason there is no center of expansion.

To illustrate this last point, a two-dimensional model is sometimes used. This is the somewhat famous balloon model, where you put a few dots on a partially inflated balloon then continue to inflate the balloon such that all the dots will move away from each other without there being a center of expansion. Very tidy but, as we will see later, this model relies on hidden structure.

In a recent publication physicist Lee Smolin (American, 1955-) stated that the BB/EI is *"an established scientific fact, which has been elaborated into a detailed story…"* and that *"we have detailed theories [BB/EI] that pass numerous observational tests."*[107] To say that BB/EI fit observations is correct, but it is correct because BB/EI were designed to fit observations, which is ad hoc fixing to the extreme and is entirely inappropriate since its only justification is to save their conjectures! This reasoning is typical of the reductionist and scientific realism agendas that are now trying to redefine the notion of a scientific theory to include such wild fantasies such as the multiverse [discussed later – page 142] .

In any case, since the 1930s, hundreds of books and thousands of technical papers have been written about the redshift, expanding universe, and the big bang. Therefore, to keep this discussion from *inflating* out of control and getting too deep into the weeds we will close with a couple of final comments.

To be sure, verification of any cosmological conjecture is always difficult, but if they rate

---

[107] Lee Smolin, "The Big Bang was the First Moment of Time, " *This Idea Must Die*: Edited by John Brockman, (Harper Perennial, New York, 2015), 32.

consideration as true scientific theories, then verification beyond mere wishful thinking and hand fixing is necessary. Unfortunately, modern cosmology seems more interested in saving the SMoC than in truth. To proclaim this pieced together patchwork, as being proven reality is a stretch, at best, and an outright counterfactual fabrication, at worse.

I believe that the expansion of the universe claim is an example of underdetermination, if not just plain miss interpretation or to paraphrase a contemporary notion – theoretical misappropriation. To end this section here is a quote that lucidly states the case: "*More specifically, our goal is to show that (1) excluding certain pathological examples, every cosmological model of our universe is empirically underdetermined;*[108] *no amount of observational data we could ever (even in principle) accumulate, can force one and only one cosmological model upon us. Additionally, we also claim that (2) even if one assumes a principle of uniformity – that the physical laws we determine locally are applicable throughout the universe – these general epistemological difficulties remain.*"[109]

Cosmology, which has somehow gained more credibility than deserved, is not a true science in

---

[108] For those who might be unfamiliar with the word underdetermination here is the wiki-def: "In the philosophy of science, underdetermination refers to situations where the evidence available is insufficient to identify which belief one should hold about that evidence."

[109] John Byron Manchak. "Can we know the global structure of spacetime?": http://philsci-archive.pitt.edu/4162/.

the sense that chemistry, biology, physics, and so on are true sciences because cosmology is strictly an observational practice devoid of experimental possibilities. Practitioners build toy models such as the well-known Friedmann–Lemaître–Robertson–Walker (FLRW)[110] model that are used to state that we know when the universe started, how it developed, and a lot of other stuff based on interpretations of observations that always lend themselves to more than one valid possibility. This is not just non-science; it is also extreme arrogance!

Adherence to the BB is clearly an example of groupthink such that these astrophysicists and cosmologists are determined to hang on to this notion and the notions of inflation and accelerating universal expansion at any cost and no matter how many hand-fixes are required. Put simply, it is likely they will continue to play around with their toy models of oversimplification and *ad hoc* patches.

Choosing a particular cosmological model is more psychological and sociological, and perhaps even political, rather than scientific. This is

---

[110] The FLRW is a toy model metric, developed in the 1920s and 1930s by the named authors, that many believe represents the structure of the universe. The metric depends on the key assumptions of universal homogeneity and isotropy in the distribution of matter (galaxies, clusters of galaxies, etc.). It also assumes that the spatial component of the metric can be time dependent. It is part of the SMoC.

partly because all models require huge extrapolations and assumptions based on observations that leave considerable room for multiple interpretations. And some of these extrapolations and assumptions have no direct observational support – or indirect support either.

There are huge differences between experimental science and observational science that put sever limits on cosmology as a true scientific discipline.

Today's science seems to be making the same mistakes made in the 18th and 19th centuries with the development of classical physics. Like those developers of classical physics, today's theorists seem to believe to soon that they have achieved their goals and cling to these theories with a religious tenacity that proclaims to have arrived at universal truths about the universe.

Multiverse and Fine-Tuning Argument

'The flights of physicists and astronomers today answer <u>the esthetic need for satisfaction of the imagination rather than to any strict demand of unemotional evidence of rational interpretation</u> [Emphasis added].'[111]

When I started writing about the BB/EI, I decided not to mention the so-called multiverse notion (or the many-worlds - or Everett – interpretation, which definitely will not be discussed).

---

[111] Dewey, (1934) 2005, 31.

However, since there seems to be growing interest in this notion about an uncountable number of universes, a few comments seemed appropriate.

The multiverse notion essentially states that our universe is only one of an innumerable number of universes, and that the one we live in was *randomly conditioned just right* for life. This notion – generally referred to as the *Fine-Tuning Argument* (hereinafter FTA) claims there is no fine-tuning (intentional or otherwise) of various constants of nature in support of life, but that it was only random chance that this universe just happened to have the correct parametric values (physical conditions) for life – lucky us?

Advocates of the FTA, like Andrei Linde (Russian-American Physicist, 1948-) claim that even though the multiverse is neither refutable nor irrefutable, it is based on the Standard Model of Cosmology (SMoC – page 134134), so therefore, it must be correct. He also stated that any argument against the multiverse must show that only one universe is possible. This is obviously backward thinking; the burden of proof is necessarily on the advocates of the multiverse, not the other way around![112]

*   *   *

---

[112] Andre Linde, "The Uniformity and Uniqueness of the Universe," *This Idea Must Die*: Ed. John Brockman, (Harper Perennial, New York, 2015), 44.

There must be some demarcation between science and non-science, and to say, as some scientists and philosophers do, that testability is naïve and amateur philosophy is not acceptable as an excuse to dismiss the notion that a true scientific theory must have true testability and make predictions not yet observed. That is, any notions that do not have these features are essentially attempts at diluting scientific evidence such that it is no longer science but fantasy. Another possibility, although weaker, is that given the correct circumstances, the ability to explain previously observed, but unexplained, anomalies might constitute a true scientific theory, if, and only if, at the same time, novel and/or testable predictions are possible.

Recently it came to my attention that in 1986 there was a Supreme Court case concerning creationism versus Darwinian style evolution. This case was settled in large part based on the written testimony *presented under the signatures of 72 Nobel Prize laureates* in support of evolution and very critical of creationism. Here is a quote from that written testimony:

"To be a legitimate scientific ''hypothesis'' an explanatory principle must be consistent with prior and present observations and <u>must remain subject to continued testing against future observations.</u> <u>An explanatory principle that by its nature cannot be tested is outside the realm of science.</u> The <u>process of continuous testing</u> leads scientists to accord a special dignity to those hypotheses that accumulate substantial observational or experimental support. Such hypotheses become known as scientific 'theories.' If a theory successfully explains a large and diverse body of facts, it is an especially 'robust' theory. If it consistently predicts new phenomena that are subsequently observed, it is an especially 'reliable' theory. Even the most robust and reliable theory, however, is tentative. A scientific theory is forever subject to reexamination and -- as in the case of Ptolemaic astronomy -- may ultimately be rejected after centuries of viability.[emphasis added]"[113]

This statement may have won the case for anti-creationists, but it appears to spell disaster for the multiverse (and the SMoC), since the multiverse notion does not meet any of the criteria ascribed by these 72 Nobel laureates, and others, since it is completely devoid of possible observation and impossible to test! If that is not convincing enough, here is what the famous physicist Richard Feynman had to say about what constitutes science: "The principle of science, the definition, almost, is the following: The test of all knowledge is experiment. Experiment is the sole judge of scientific 'truth' [emphasis in the original]."[114] Again, here is a strong statement of the

---

[113] Edwards v. Aguillard: U.S. Supreme Court Decision. Brief submitted under the signatures of 72 Nobel Laureates, 17 State Academies of Science, and 7 other scientific organizations in support of the appellees. www.talkorigins.org/faqs/edwards-v-aguillard/amicus1.html

[114] Richard Feynman, *The Feynman Lectures on Physics: Volume I*. (Addison-Wesley, Reading MA, 1989), 1-1.

indispensable requirement of testability if a notion is to be considered scientific.

As another example, according to Bernard Carr (British astronomer, 1948 -), as historical beliefs about the design of the universe went from the claim of the universe being geocentric to the later claim of being heliocentricity, the multiverse is just the next step in the process.[115] However, there is, at least, one major problem with his statement, it ignores the fact that the other changes mentioned were due to improved instrumentation and observation, while the multiverse has no observational support regardless of instrumentation, or any basis that can honestly be considered a fact. Obviously, the multiverse conjecture is a notion that can never be proven or disproven through observation or any other means.

True advances in science have not been made by modeling data, but by intuitive thinking followed by conjecture with subsequent attempts to verify the predictive and explanatory capabilities of the conjecture through experimental testing. The multiverse meets none of these criteria and has no predictive capability and by allowing any possibility, it really does not explain anything!

Linde and the other advocates of this fantasy are correct when he says that the multiverse notion is irrefutable, so is the existence of the *tooth fairy*. It

---

[115] Bernard Carr, as quoted in Tim Folger, 'Science's Alternative to an Intelligent Creator: The Multiverse Theory', *Discover Magazine* (December 2008), online version.

is also non-measurable in the sense of observational confirmation or any empirical verification at all. The multiverse and BB/EI together amount to a *theory of anything and to predict anything is to predict nothing*!

Put another way, the multiverse/FTA argument essentially states that the physical parameters that govern the physics of the universe were randomly (accidentally?) fine-tuned for life, and, as discussed above, this is claimed to imply the existence of an almost innumerable number of other universes, each with different parameters and parametric values that are *somehow* randomly selected. These parameters are the 26 or so constants of nature that are part of the Standard Model of Particle Physics (page 78).

Arguments for fine-tuning claim that the existence of a universe that supports life is a probabilistic issue and that as stated by philosophy professor Peter Epstein (UC Berkeley) "…conditions needed to support life are extremely unlikely to occur in <u>any given universe</u> according to our best cosmological theories [Emphasis added]."[116] Since it is impossible to know anything at all about *any given universe*, we can only know about this universe such that making such statements about any other universe is essentially fanciful

---

[116] P. F. Epstein. "The Fine -Tuning Argument and the Requirement of Total Evidence", *Philosophy of Science Journal Volume 84*, 2017, 640.

wishful thinking and is certainly not science or even good philosophy!

On the same page, Epstein goes on to say; "…the chances of there being some life supporting universe are higher the more universes there are…" How can someone make such statements when 1) there is absolutely no substantive evidence that even remotely suggest the existence of other universes, 2) This statement pre-supposes a knowledge of what constitute a life viable universe about which we know almost nothing beyond speculative imagination, 3) his argument about the more universes the more likely to have at least one supporting life is a perfect example of the Gambler's Fallacy [discussed below]. The rest of Epstein's article is essentially a continuation of unsubstantiated and unverifiable statements about probabilities, likelihoods, and so on. It seems that Epstein, like many others, is willing to go to any extremes of imagination to avoid anything that might even hint at purpose or intentionality within the universe.

Furthermore, with regards to these physical parameters that are claimed to determine the structure and, indeed, the very existence of the universe, I believe that it can safely be said that no one knows enough about the interaction of these parameters to really make claims about the possibility of being compatible with life. In support of that statement, here is a quote from a recent (February 2019) paper that discusses both the multiverse and the fine-tuning argument:

"In particular, the range of possible parameter values (the minima and maxima of the distributions) must be specified. A full assessment of fine-tuning requires knowledge of these fundamental probability distributions, one for each parameter of interest (although they are not necessarily independent). Unfortunately, these probability distributions are not available at the present time [emphasis added]."[117]

Put simply, this means that permissible parameter values are not known, and how these parameters may vary and interact is not known, so any attempts to model possible universes by manipulating these parameters is simply guess work.

Wait, there is more, this review goes on to admit that the multiverse notion is simply a fantasy:

"The consideration of possible alternate universes, here with different incarnations of the laws of physics, is by definition a counterfactual enterprise. This review considers the ranges of physical parameters that allow such a universe to be viable. Since alternate universes are not observable, this endeavor necessarily lies near the boundary of science. Nonetheless, this discussion is useful on several fronts: First, one can take the existence of the multiverse seriously, so that other universes are considered to actually exist, and the question of their possible habitability is relevant [emphasis added]."[118]

---

[117] Fred C. Adams. "The Degree of Fine-Tuning in our Universe – and Others", https://arxiv.org/abs/1902.03928 (Feb 2019), 12.

[118] Ibid, 12.

Here is a candid admission that the multiverse notion is pure fiction and fantasy. Again, put simply, the notions put forward concerning the multiverse are simply fictitious and contrary to the second emphasized statement that the multiverse notion, is not just near the boundary of science, but does indeed lie outside the boundary of science, and inside the boundary of fantasy, and the last statements underlined in the quote is really amazing since it states that even though the multiverse notion is make-believe, *we can pretend that it is fact*! This paper continues this fantasy narrative for more than <u>200</u> pages pretending that something significant could actually come out of this fictional story.

Some attempts at explaining fine-tuning apparently try to analogize with some type of random process, such as a giant billion slot roulette wheel, catching the largest fish in a lake, or some other mechanism of apparent random selection that gets it just right. These arguments are simply weak attempts to explain in reductionist terms how our universe just happened to support life. Also, they never seem to explain who or what spins the roulette wheel or catches the fish.

On the other hand, it is surprising that sometimes we see references that imply purposeful tuning from those who usually denounce such notions. Here are two examples from well-known scientists – Lee Smolin and Stephen Hawking:

"The existence of stars rests on several delicate balances between the different forces in nature. These require that the parameters that govern how strongly these forces act be <u>tuned just so</u>."[119][Emphasis added]

"The remarkable fact is that the values of these [fundamental] numbers seem to have been <u>very finely adjusted</u> to make possible the development of life."[120][Emphasis added]

These two phrases *tuned just so* and *very finely adjusted* certainly seem to imply a non-random process, which, in turn, implies purposeful and intentional action – however, no suggestions seem to have been made about who or what did the fine-tuning.

As mentioned above, some of these arguments are victims of the *Gambler's Fallacy* and others the *Inverse Gambler's Fallacy*. The Gamblers Fallacy, for example, is the assumption that a after a long sequential series of *heads* appearing with the flip of a coin, that the odds improve for getting *tails* on the next toss. When, in fact, the probability does not change for the next flip. On the other hand, the Inverse Gamblers Fallacy is the mistake of inferring a large number of trials (such as a lifetime's worth of poker hands) solely from observation of a single success (such as a Royal Flush). Winning a lottery, does not mean that

---

[119] Lee Smolin. *The Life of the Cosmos*. N Y, NY: Oxford UP, 1997. 37.

[120] Stephen Hawking. *A Brief History of Time: From the Big Bang to Black Holes*. New York, NY: Bantam Books, 1988, 125.

person has played the lottery many times, it could just as well be their first try.

Arguments about fine-tuning seem, for the most part, to miss the point. Life exists so quibbling about whether it is rare, unique, or abundant is essentially a philosophical question, not a scientific question. It is also questionable whether any discussion about fine-tuning should only refer to micro-features defined in the Standard Model of Particle Physics.

Instead of asking why the *laws* of nature seem so *fine-tuned* for the support of life, perhaps a better question is how did life managed to conform to the existing laws of nature? That is, it seems that life is fine-tuned for compatibility with the laws of nature, not the other way around. Since the environment existed before life, it is trivially obvious that life found a way to exist within this environment not the other way around.

There is no doubt that this universe does provide conditions compatible with some type of life, and that life found a way to exist and thrive within this planet's environmental conditions. By extension, it seems reasonable to assume that since life can and did occur here on earth, then life can and did occur on a myriad of other planets as well.

It seems strange and hard to understand how reputable scientists can state, seriously, that the FTA or any other argument confirms the multiverse notion. Perhaps, as I personally believe, it is because of hidden agendas that these scientists support the FTA and the multiverse notions as a means of suppressing any non-materialistic beliefs.

As discussed, there are sound rationally logical and physical reasons to reject both the notion of the multiverse and the FTA, while, conversely, there are no sound rationally logical reasons to accept them. Given this dichotomy, it is clear that the multiverse and the FTA should be relegated to the world of science fiction and fantasy.

As a final comment on this it should be noted that some of these *reputable* scientists are using US taxpayer money from the National Science Foundation and other tax-based sources to promulgate these fictitious notions under the guise as being scientific!

Universal Motion

"The experiences which prove the dynamics in-equivalence of different states of motion teach us that the world bears a structure. However, in the concept of absolute space this inertial structure is evidently not sized up correctly; the dividing line does not lie between rest and motion but between uniform translation and accelerated motion. [Emphasis added]"[121]

As this quote implies, all matter is in constant motion and there is only the question of whether

---

[121] Hermann Weyl, *Philosophy of Mathematics and Natural Science* (Princeton, NJ: Princeton University Press, 1949), 101.

this motion is uniform translational motion or accelerated motion.

Although the terms uniform translational motion or accelerated motion may seem unfamiliar, we have all experienced motion of both types. Uniform translational motion is experienced when riding in a vehicle that is moving at a constant velocity. More technically, an object moving with uniform translational motion is said to have an *inertial frame of reference*. That is, using the example of an automobile moving at constant speed, the motion of the vehicle is only known by observing the apparent motion of objects outside of the vehicle.

On the other hand, it is only through accelerated motion that there is anything like a *non-inertial frame of reference* and that is the system referencing the body being accelerated (linear or rotation). Put another way, accelerated motion has an absolute aspect as *felt* by the object being accelerated.

Although a change in velocity may be observed by an observer outside of the moving object, the absolute aspect of acceleration is only applicable to the body being accelerated, not to any observer of that acceleration, who only see an increase in separation and rate of separation. That is, accelerated motion may be observed through change in velocity with regard to the observer, but only the object being accelerated *feels* the acceleration, which is due to inertial mass resisting the accelerating force.[122]

Accelerated motion comes in two types. The first is the linear acceleration that we experience when a motor vehicle accelerates in a straight line. The second type is angular acceleration, which is experienced when a vehicle is turning a corner.[123] Put another way, the mass of an object in uniform translational motion feels no resistance to its motion, while, on the other hand, an accelerated object *feels* a force proportional to the object's mass and the force causing the accelerated motion.[124] As mentioned above, uniform motion is only detectable through observations of a *non-comoving frame of reference*, such as the view of objects outside a moving vehicle. Accelerated motion is self-realizing within the objects coordinate system.

For humans, the feel of acceleration is partly due to anatomical organs that are sensitive to changes

---

[122] Of course, the same can be said about decreasing relative speeds where an object experiences deceleration, the reverse of acceleration. Both types are familiar to anyone who has traveled in a commercial airplane where the acceleration at take-off and the deceleration on landing are very apparent.

[123] This angular acceleration is independent of whether or not the vehicle is moving at a constant speed. For example, an object rotating about an internal axis experiences rotational acceleration such as a spinning top or the daily rotation of the earth.

[124] That is, force is equal to mass multiplied by acceleration ($F = ma$). Or, put another way, acceleration (a) equals the applied force (F - such as gravity) divided by the mass (m) of the object being accelerated, or $a = F/m$.

in our motion relative to the earth's surface. Put another way, this sensation of acceleration that we commonly feel is due to the inertial effect on matter when the rate of motion changes relative to what was, until the change in velocity, a co-moving or inertial frame of reference – the earth.

This brings us to the topic of this section, which is universal motion, or, simply, the notion that all matter (as defined on page 9) is always in relative motion – either uniform translational or acceler-ated (linear or angular). If this notion of constant or universal motion seems an overstatement, then consider the reverse and ask what matter is not in motion and when is this matter motionless.

To be sure, whether it is the constant orbital and rotational motion of our planet or the motion of other celestial bodies, all earth-bound matter is always in relative motion. Obvious examples in-clude. the apparent daily motion of the sun and the nightly apparent rotation of the stars (except one). More subtle examples include the motion within a glass of water that has set undisturbed for some time. The water seems perfectly still, but we know that the water molecules are constantly bouncing around in a motion called Brownian movement.[125]

---

[125] This phenomenon was discovered by Robert Brown (Scottish, 1773-1858) in 1827 while experimenting with some pollen sus-pended in water. It clearly demonstrated that the molecules of water were in constant motion even when left undisturbed.

Although. At the level of everyday experience, other agents play a role, the dominant agent, for the universe as a whole, is gravity. Even though gravity is many orders of magnitude weaker than other forces such as electromagnetism, gravity has the overwhelming advantage that it is pervasive and cannot be blocked or cancelled.[126]

When considering concepts such as mass, inertia, and motion, it should be kept in mind that there are no physical objects that are entirely force-free. Only idealized mathematical abstraction or other mental constructions can be said to be force-free. Objects in a state of equilibrium regarding its local frame of reference are in this state by balanced forces acting on that object. For example, the rock setting on the ground pushes against the earth with a force that is equal to the force that the earth is pushing against the rock.

One can imagine gravity as that persistent and irresistible force that is always active and from which there is no escape. The weightlessness or *free-fall*, as experienced by orbiting astronauts, is not free. It is a delicate balance and constant struggle between gravity and the angular acceleration of the orbiting object. Gravity will not go away!

---

[126] The gravitational constant is 10^{-36} (= 0.000000000000000000000000000000000001) times weaker than the electromagnetic constant!

Side Comment on Gravity: Another interesting fact about gravity is that all objects accelerate at the same rate if other factors such as air resistance is eliminated such as in a vacuum, where a rock and a feather will fall at the same rate. It is indeed quite remarkable that, in a vacuum, all objects fall at the same rate– that is with the same acceleration in a manner that has yet to be explained. Any physical body in the same gravitational field will have the same acceleration such that it is not the objects mass that determines the rate of acceleration but only the fact it is a material object of some kind with properties that equates the rate of acceleration for all objects. We know a lot about gravity, but we do not know how or why this equal acceleration works.

Side Comment on Universal Change: An adjunct to, and concomitant with, universal motion is universal change. There is a continuous and irresistible cycle from undifferentiated substance into differentiated substance and back again into undifferentiated substance. Universal change is an enfolding and unfolding, nothing remains unaltered. For example, undifferentiated non-living substances are enfolded as differentiated living organisms, then eventually unfolded back into undifferentiated substances.

Discrete versus Continuous Motion
Given universal motion, there remains the question of whether this motion is discrete or continuous? Like other questions examined herein, this is an ancient question, whose origin

is lost in antiquity. However, we do know that it dates at least as far back as ancient Greece and most notably to the paradox-like scenarios attributed to Zeno of Elea (490–430 BC).

Zeno's paradoxes rely on the notion that an object would have to pass through an infinite number of positions when moving from some position *A* to another position *B*. Zeno speculated that movement through an infinite number of intermediate positions would require an infinite amount of time and, therefore, motion does not exist and is an illusion.

One of the problems with this paradox of infinite positions requiring infinite time is the category mistake of applying *physical reality* to an *abstract mathematical concept*. A category mistake is something like comparing apples and oranges, or, since apples and oranges are both fruit, more like comparing apples and rocks. In this case, the category mistake is improper comparison of an assumed physical continuum (infinite number of physical positions) and an abstract continuum (mathematical infinity).[127] Or put another way, the mistake of equating a mathematical abstraction (infinite number of points on a line interval) with a physical length or distance between two objects is an invalid comparison.

---

[127] Within mathematics, there is something called the *continuum of numbers,* which, put simply, is the question of how many points are on a Euclidean line of any finite length.

It is sometimes said that Zeno's paradox is solved through a finite sum of an infinite number of terms. This makes a good point, however; this requires that the infinite series of terms converge to a finite number, which is not always the case.[128]

Fast forward about 2400 years and it turns out that all this is connected to the so-called *quantum of action* or Planck's constant (usually designated by the letter h). This was the discovery of the physicist Max Planck (German, 1858-1947) who, in 1900, published a paper demonstrating that at the microscopic level energy was incremental. This level, called the Planck scale, is too small for direct probing and is so removed from phenomenal experience that it has been said that physics meets philosophy at the Planck scale.

According to the Planck's theory, the smallest possible increments of energy are proportional to his quantum of action and, by implication, the Planck scale represents the smallest possible increments allowed by nature. This scale includes values, representing time (in the sense of event separation), length, and mass (energy), which respectively represent the smallest possible increments – for each type of measure. This means that at the Planck scale motion is

---

[128] It should be noted that the sum of an infinite series is never calculated using an actual infinity of terms but approximates an infinite number of terms through algebraic manipulations. For example, the repeated tossing of a coin can be equated to an *infinite* series where the percentage of heads (or tails) asymptotically approaches 0.5 or 50%.

incremental. This, in turn, means that between any two physical positions there is always a finite number of increments.

OK, maybe this seems to be stretching it a bit. The motion we commonly experience certainly seems continuous and smooth, but that is just because of our perceptual capabilities. Obviously, we cannot view motion at the Planck scale, just as we cannot see the individual molecules of flowing water or the discrete molecules of any surface that appears both smooth and continuous. For example, the paper that I am now writing on feels smoothly connected but if I were to look at it through a microscope it would appear extremely lumpy and rough.

So, what is the point? The point is that, as with other notions discussed herein, a deeper understanding of our natural world increases our appreciation of the wonders of nature and enhances our worldview. Although, not commonly important in everyday situations, discrete versus continuous motion is important when scientifically or philosophically contemplating the fundamentals of nature.

When Planck developed the notion of the quantum, it was a leap of imagination that shook the scientific world at the most fundamental level. More than just a paradigm shift, this notion that nature, at its most fundamental level, was discontinuous and not continuous was truly revolutionary and counterintuitive to an extreme. It was difficult enough to consider the notions that energy at the smallest level was discontinuous, but the that matter, motion, and light were also discontinuous was even more difficult to accept.

This notion came at a time when the atomic nature of matter was still being debated and definitely made the atomists position much stronger, but the *continuous faction* was still not finished off.[129] That finish came a few years later by two papers that Einstein published in 1905, the one on Brownian movement and the one on the photo-electric effect. The first, convinced most skeptics that matter was atomic in nature, while the second reinforced Planck notion of energy packets and, although not realized at that time, was the beginning of learning about the interactions between photons and electrons that eventually grew into the theory of quantum electrodynamics.

Because of Planck's discovery, the year 1900 might well be considered a point of demarcation between the classical scientific world and the new physics that turned classical science on its head. Unexpected discoveries such as the constancy of the speed of light, the strangeness of black body radiation, and the new discovery of the abnormal orbit of the planet Mercury were mysteries that needed solving. It turned out that these three anomalies were resolved by Max Planck, Albert

---

[129] Four years earlier when radioactivity was discovered the position of the atomists was all but confirmed.

Einstein, and others over the next twenty or so years. However, science was to pay a price for these discoveries in the form of new and strange paradigm changes that even today continue to confound intuitive thinking about the natural world.

Side Comment on Continuity versus Continuous: Nature has continuity, but nature is not continuous. Discrete events, such as the scenes in a cinema that are systematically ordered have continuity, but they are not continuous in the sense of having no gaps or interruptions. Continuity does not, necessarily, mean seamless connectedness (continuous), but is a reference to a certain sense of order.

Side Comment on the Planck Scale: Currently, particle accelerators probe into the micro-world at the level of about 10^{-17} centimeters, which is about half way to the ultimate smallness of the Planck length of about 10^{-33} centimeters. As pointed out by D. Bohm (American physicist, 1917-1992)[130], we should consider that since the orders of magnitude of complexity between the macro-level of human experience and the smallest level yet probed by science is only about half way to the Planck level, this leaves plenty of room for a great deal of undiscovered complexity and hidden reality.

---

[130] David Bohm, *The Undivided Universe*, (Routledge, New York, 1993), 38.

The Infinite

Infinite (or infinity) is a word that has been much used, and much abused to the extent that most usage of this word is either imprecise or just plain wrong. Moreover, it is safe to say that no one who uses this word has ever experienced the infinite in any form. In fact, the infinite, if it exists in any form experienced by humans, exists only as an abstract mathematical convention that was mentally conceived as a calculational convenience.

Infinity is a label we attach to the notion of something increasing or decreasing without limit. That is, the term infinite is not only used for something increasing without limit (an exterior infinity – the infinitely large such as some believe the universe to be), but also, as Zeno apparently assumed, for something that decreases without limit (an interior infinity – the infinitely small such as mathematical notions about the points on a line). Surely, such a concept lies beyond the limits of ponderable comprehension and defies intuitive understanding. The truth is that we are able to deal with the notion of infinity only through non-intuitive symbolic manipulation.

Without doubt, it would be better if most of us just did not use the word infinite and find a less imponderable word (interminable, limitless, really big or small?) because what appears to be extension without limit could be just going around in circles.

The Void

"It is impossible to picture empty space. ... <u>Whoever speaks of absolute space uses a word devoid of meaning [Emphasis added]</u>. This is a truth that has been long proclaimed by all who have reflected on the question, but one which we are too often inclined to forget."[131]

"It will be seen that though geometry is not an experimental science, it is a science born in connexion with experience; that we have created the space it studies ... <u>We have chosen the most convenient space, but experience guided our choice</u>. As the choice was unconscious, it appears to be imposed upon us. Some say that it is imposed by experience, and others that we are born with our space ready-made [Emphasis added]."[132]

"To be material is to have a spatial position. The number 2, for example, has no spatial position. Spatially contiguous objects represent one another. We do not know what the objects are if we know merely that they are in space. <u>We understand space only through the drive of the objects in space; otherwise we have no idea what space is</u> [Emphasis added]." [133]

"Let us then see what we are thinking about when we speak of 'Nothing.' <u>To represent 'Nothing,' we must</u>

---

[131] Henri Poincaré, Science and Method, Trans: Francis Maitland, (London: Thomas Nelson and Sons, 1918), 93.

[132] Ibid., **115.**

[133] Kurt Gödel Quoted in *A Logical Journey*, Hao Wang (Cambridge MA: MIT Press, 1996), 292.

<u>either imagine it or conceive</u> it. Let us examine what this image or this idea may be [Emphasis added]." [134]

The first quote questions the validity of absolute or physical space, while the second quote offers a choice between learned space and instinctive space. Herein, we propose that we learn the concept of space at an early age and it is deeply ingrained such that it has the appearance of being intuitive. That is, because our senses limit us to a three-dimensional world, we assume that we are embedded within a three-dimensional space, although there is no substantive evidence to support this belief.

To be sure, rooting out this notion of three-dimensional space, or any other, ingrained notion is difficult even when directly confronted with a viable alternative. The third quote suggests that it is only through positional relations between objects that we build our notions about space, and, as the fourth quote states, visualizing nothing is not so easy and perhaps, all we can do is define it.

Before continuing, I would like to list a few quotes that I gleaned from a well-known online philosophy forum about the believed nature of space:

"<u>We know that space has properties, so it is real</u>. One particular property prevents light from propagating at infinite speed [Emphasis added]." – *Does this person*

---

[134] Bergson 1911, 293.

really believe that it is a property of space that restrict the speed of light? Also, I would like to see a list of the properties of space.

"Physical space is then a container of objects, shared and common to all observers. This space only differs for each observer by their respective metrics [Emphasis added]." – *So, for this person space is a box – then where are its boundaries and what is this box made of?*

"It's better, I think, to view the relationship between matter/energy and space as one of coinciding or coincidence: particles or fields of matter or energy are said to coincide with parts of space. Rather than just a passive container, space has the power to structure matter in three and only three dimensions."[Emphasis added] – *For this person, space is the source of three-dimensional matter!*

I can only say that these three opinions by academic philosophers seem somewhat naïve and I find their notions disturbing for their lack of critical thought and insight. I suspect that these quotes are partly the result of mis-interpreting the theory of general relativity with the mistaken belief that there is a physical relation between space and gravity – this common mistake is discussed in the next section – *General Relativity and the Void*. Anyway, to continue…

If we remove all matter, what is left? Put simply, without matter there is nothing. As implied in the third quote- at the beginning of this section – the term space only has meaning within the context of matter. It is from the relational intervals between substantive objects, that we build our notion of space. Material objects possess physical reality and a relational position, but we are mistaken when we regard the relational connections between these objects as residing in some physical container. Every reference to an object requires one or more other objects in order to establish positional relationships – there obviously being no relationship of just one.

As the fourth of the above quotes reminds, intuitively we want space to be some-thing because we cannot directly experience nothing and must imagine it through thought.

After some thought, it seems truthfully apparent that without matter, there is only the Void. It is incorrect to speak of *empty space*; empty implies a container of some sort. The Void – being nothing – has no dimensions, no extension; it is total absence. As nothing, the Void cannot be the cause of anything. If this is so, what then is the cause of our belief in three-dimensional space? This question will be answered later.

Side Comment on Giordano Bruno: In the year 1600, Giordano Bruno, a Dominican Friar, was burned at the stake partly for his views on cosmology and partly on his lack of piety (to put it mildly). Bruno was more a philosopher than a scientist, but among other things, he heretically proposed that the stars were actually distant suns and that the universe contained innumerable inhabited worlds! For these and other beliefs, he was condemned by the Inquisition.

To see why this happened, it is helpful to take a brief look at some of his ideas that were published over the 20-year period prior to his demise. Although, some or even most of his ideas were not original, his manner of presentation was original. Bruno postulated that other than the stuff of the universe, there existed only the Void. For example, here are a couple of quotes from his most well-known work:[135]

"Sixthly, if we posit a finite world, it is impossible to escape acceptance of the void, if void is that which containeth naught."

"Seventhly, this space in which is our world would without it be indeed a void, since where the world is not, there we must infer a void."

Additionally, Bruno conjectured that earth is just one of an infinite number of worlds and he differentiated between the planets – that reflected light and the suns – that produce light.

"…it is shown that although some bodies are luminous and hot of their own nature, yet it doth not follow that the sun illumineth the sun and the earth illumineth herself, or that water doth illumine itself. But light proceedeth always from the opposed star…"

"…but those suns, bodies in which fire doth predominate, move differently to the earths in which water predominateth; thus may be understood whence is derived the light diffused by stars, of which some glow of themselves and others by reflection."

---

[135] Giordano Bruno. *On the Infinite Universe and Worlds*, Venice, 1584.

He also believed that many other earths exist that have life and the glory of God has been made manifest in these worlds as well.

"...in which are all those worlds which contain animals and inhabitants no less than can our own earth, since those worlds have no less virtue nor a nature different from that of our earth."

"...Thus, is the excellence of God magnified and the greatness of his kingdom made manifest; he is glorified not in one, but in countless suns; not in a single earth, a single world, but in a thousand-thousand, I say in an infinity of worlds."

That is quite a list and after making the mistake of moving back to Venice, he was accused of heresy and turned over to the Roman Inquisition. He was taken to Rome and there he remained several years in a Roman jail where every now and then he was brought out for questioning and then returned. Exactly what charges were brought against Bruno is unknown but eventually, in 1600, he was finally condemned and burned at the stake at a piazza where now stands his statue.[136]

General Relativity and the Void
The commonly accepted interpretation of general relativity states that space is bent and distorted under the gravitational influence of large bodies (planets, stars, etc.). This notion that space is curved is based on an interpretation of the

---

[136] The statue is in the Roman piazza *Campo de' Fiori*.

mathematics of general relativity that tacitly assumes the existence of absolute space.

That is, the mathematics of general relativity can be interpreted such that space is absolute in the sense that space is something that can be bent or otherwise distorted. If space were a container filled with some kind of substance, then according to this common interpretation of the equations of general relativity, gravitational forces either cause the assumed three-dimensional space to bend into another dimension or there is stress without bending.[137]

Traditional discussions about general relativity often use models such as that of a bowling ball rolling around on a trampoline. This is supposedly analogous to the impact that large masses (planets and stars) have on three-dimensional space. In this model, the two-dimensional surface of a trampoline is distorted into a third dimension.

Since it is best to go right to the source, here is what Einstein had to say about general relativity back in 1915 when it was first published:[138]

*"We introduce in a space, which is free from Gravitation-field, a Galilean Co-ordinate System **K**(x, y, z, t)*

---

[137] The American physicist John Wheeler was wrong when he made this well-known statement: '*Spacetime tells matter how to move; matter tells spacetime how to curve.*'

[138] In his later years, Einstein made the comment that after the mathematicians had taken hold of his theories, he no longer recognized them.

*and also, another system **K'**(x', y', z', t') rotating uniformly relative to K. The origin of both the systems as well as their Z-axes might continue to coincide. We will show that for a space-time measurement in the system **K'**, <u>the above established rules for the physical significance of time and space cannot be maintained</u> [Emphasis added].*"[139]

"Because among all substitutions there are, in every case, contained those, which correspond to all relative motions of the co-ordinate system (in three dimensions). <u>This condition of general covariance which takes away the last remnants of physical objectivity from space and time, is a natural requirement, as seen from the following considerations</u> [Emphasis added]."[140]

Both of these quotes clearly state Einstein's position on the reality of space and time. Without troubling about the **K** and **K'** stuff in the first sentence, the first quote clearly states that maintaining any notion of a physical space-time is untenable, while the underlined sentence in the second quote clearly states that the removal of the last remnants of physical objectivity from space and time means that both space and time are mental constructions. These notions are both

---

[139] A. Einstein, *The Foundation of the Generalised Theory of Relativity (1915)*:
https://en.wikisource.org/wiki/The_Foundation_of_the_Generalised_Theory_of_Relativity#C._The_Theory_of_the_Gravitation-Field, 5.

[140] Ibid, 6.

compatible and supportive of the concept of the Void and the non-reality of time.

That is, as presented by Einstein, and in agreement with the Void, there is no space to distort. The stress between the large bodies is a gravitational struggle, which only gives the appearance of a distortion of space. The physical effects of this apparent distortion results from *gravitationally defined trajectories* that behave exactly as curved space would, if space were some-thing that could be curved.

In the next section of this original paper, Einstein goes on to mention that:

"...it is clear, that from the physical stand-point the quantities g_{ab} indices are to be looked upon as <u>magnitudes which describe the gravitation-field with reference to the chosen system of axes</u> [Emphasis added] ..."[141]

Here the underlined statement clearly states that the geodesics (the shortest path between two points on a sphere or in this case the shortest distance between any two extraterrestrial bodies), are proportional to the tensor quantities g_{ab}, which represent *gravitational curvature* and not space curvature.

That is, the apparent curvature that is generally attributed to space, is instead the curvilinear path dictated by gravity and the relative movement of

---

[141] Ibid, 7. (The g_{ab} *are the components of the basic metric tensor.*)

bodies. For example, interplanetary probes do not follow the curvature of space; instead they follow complex trajectories defined by gravitational fields (plus other factors, such as the relative motion of these massive bodies).

Mathematical equations stand alone as syntactical statements that are devoid of semantics (content). These equations always are subject to multiple possibilities for interpretative content.[142] Agreement between the equations and the theory is necessary, but not sufficient. That is, the interpretation of the equations must be made in a wider context than just the content of the theory. For example, confirmation through future observation, that is, a phenomenon not previously observed that lends credence to the theory (predictability), or some experimental confirmation that increases theory confidence.

Put another way, the mathematics of general relativity, as with all mathematical equations, only describes behavior and form, not content, and through the introduction of content these equations are subject to different interpretations. Semantic free equations are mute when it comes to interpretations since mathematics is concerned with relations between conceptual structure, not with interpretive context.

---

[142] The Inverse Square Law equations have the same form but has multiple interpretations with the application of context such as the mass of two bodies for gravity or the intensity of two electrostatic charges for Coulomb's Law.

For example, the equations for geodesics may be erroneously interpreted as curvature of space or, more correctly, as gravitational stress between massive bodies such as planets, stars, and galaxies. Continual thinking in terms space and time as having some kind of physical existence is an obstacle to clear fundamental thinking about the interactions of physical bodies and the structure of the universe on the broadest scale.

It is unfortunate that after more than one hundred years, references to physical space-time are still to be found in technical papers – although as originally put forward –Einstein's theories of relativity clearly rejected such notions.

Dimensionality

After the previous discussions, it may seem strange that the next topic should be dimensionality. This is because we commonly think of dimensions as dimensions of space, but that is just because that is what we call them, and we are accustomed to thinking that way. To discuss *spatial* dimensions within the context of the Void is inappropriate. On the other hand, inasmuch as we commonly experience three dimensions, dimensionality is part of our reality.

It appears that many, if not most, arguments in favor of a three-dimensional universe assume that three is the correct number of dimensions and then proceed to defend this choice. In most of these arguments, three-dimensional space seems based on the assumption that, as discussed in the prior section on the Void, there must be some kind of container or at least an entity capable of exhibiting some number of dimensions. It seems that we naturally like things tidy and in boxes – in this case a three-dimensional box. However, since the Void is total absence, the Void has no dimensionality. If that is so, then why do we perceive a three-dimensional world?

The simple answer is that we experience a three-dimensional world because our senses are apparently only capable of building up a perceptual space of no more than three dimensions – beyond three dimensions we lose our intuition and ability to visualize objects of more than three dimensions. Our visual space is based on two-dimensional perception, which, as discussed earlier, is processed through accommodation and correlation into three-dimensional images that are then mentally interpreted as objects in three-dimensional space.

This notion, of three-dimensional space, is also supported by our tactile space, which is constructed from our ability to feel objects only in three dimensions. In this way, the idea of three-dimensional space is built up through the limitation of our senses and the lack of mental capability to intuit or visualize beyond three dimensions.

Of course, these limitations do not preclude the possibility of hidden structure.

Hidden Structure Model

First, this model is not a representation of space, but is a geometric exercise that illustrates how, even in a simple model, structure can be hidden.

Consider a loop such as a circle or any closed figure that does not cross over itself (like *0*, not like *8*). This type of loop allows a moving point unrestricted movement, such that a moving point will eventually return to where it started. For example, visualize a coffee cup with a bug walking around the rim, this bug has unrestricted movement, but only in one dimension. Next, consider the surface of an inflated balloon. If a bug is walking on the surface of the balloon, the bug will have unrestricted movement in two dimensions.

However, in both examples, unseen from the bug's point of view, there is hidden structure. The one-dimensional loop (or coffee cup rim) is embedded on a two-dimensional surface. The two-dimensional surface of the balloon also requires hidden structure, since the surface of the balloon is a two-dimensional surface embedded on a three-dimensional sphere.

Surface geometry deals with the outer measurement (cup rim, balloon surface), which gives the appearance of being independent of the geometry in which the surface is embedded, but the fact that there is unrestricted movement in both models relies on hidden structure. It should be noted that these examples may be extended into more dimensions and the hidden structure require-ment remains.[143]

Under certain circumstances, there are hints or clues that reveal possibilities about this hidden structure. For example, unusual occurrences, sightings, and what might be called non-halluci-natory visions are in some cases possibly a result of hidden structure. For example, there is the oc-currence where one sometimes has the strange feeling of being watched and on looking around sees that this is so. Another example is the phe-nomenon called synchronicity. This term, coined by the eminent psychologist Carl Jung (Ger-man/Swiss, 1875-1951), refers to the synchronized occurrence of seemingly uncon-nected events and is briefly discussed next.

Synchronicity
There exists a certain type of phenomenal events that many have experienced, but few have con-sidered beyond thinking them only coincidental. This is a reference to those occurrences, of two or more events that pose meaningful connections,

---

[143] In the technical literature, seeing this hidden structure is some-times referred to as *God's view,* with the obvious implication that, like the bug, such hidden structure is unvisible to us but is observable only by Deity. On a personal note, some years back, I emailed the well-known Stanford physicist Leonard Susskind about hidden structure and his polite response was that I should consider it as analogous to the scaffolding used in constructing buildings. A nice response, but I cannot help feeling that there is more to hidden structure than that.

but appear as an effect without a cause or process. That is, they are related by meaning, but not through any known causal connections.

In the 1920s, the eminent psychologist, C. G. Jung (German/Swiss, 1875-1961) studied this type of occurrence and coined the term synchronicity.[144] Because of rareness, lack of ability to reproduced, and a lack of causal evidence these events fall outside of our normal understanding of cause and effect. Therefore, whether these events occur in a purely random manner or whether they are the act of some unknown influences has been impossible to determine.

According to Jung, synchronicity is a principle of coincidental events connected by simultaneity and meaning. This implies that besides what we recognize as cause and effect, there are possibly other factors within nature that express themselves in the arrangement of events and appears to us as meaningful - not forgetting that *to be meaningful* is an anthropic (human) interpretation.

One of the problems with synchronicity is that our sense of reality requires a postulated uniformity with known, or, at least, plausible cause and effect. Normal sequences of cause and effect create a comfort zone, while synchronistic

---

[144] Carl Jung, *Synchronicity: An Acausal Connecting Principle* (Princeton and Oxford: Princeton University Press, 1960).

anomalies disrupt this sequence and lie outside our comfort zone. Put another way, synchronistic events are connected through meaning not through any known cause-effect and usually they seem incapable of causal connection.

However, one should keep in mind that lack of understanding does not negate the possibility of the existence of complex processes whose mechanisms are presently unknown. In any case, it is not possible, now, to decide with certainty whether these unique and irreproducible events are the result of hidden causes or are merely random events lying beyond the fringe of statistical predictability.

Additionally, synchronicity should not be confused with apophenia or serendipity. Apophenia is the seeking of patterns or connections in what is random or meaningless data. For example, seeing figures in the clouds and other attempts at finding hidden patterns. Apophenia, in a certain sense, might be considered as an over-active natural inclination to find meaning. Just as we naturally construct concepts, seeking patterns and meaningful content is a natural human trait. On the other hand, serendipity refers to chance occurrences that are generally favorable such as an unexpected pleasant event or discovery. For example, the fortuitous discovery of something not looked for.

Dabbling in Higher Dimensions

"Quantum physics in particular seems to indicate that physical reality is something still more different from the appearances than even the 4-dimensional Minkowskian world. T. Kaluza's fifth dimension points in the same way."[145]

In 1909, the physicist Gunnar Nordstrom (Finnish physicist, 1881-1923) speculated that the universe might include a dimensionality higher than the three dimensions we normally experience. Unfortunately, Nordstrom published his ideas in an obscure journal and, consequently, his ideas received little attention and were soon forgotten. Following Nordstrom, there are the better-known efforts of Theodor Kaluza (German mathematician, 1885-1954) who first published on extra dimensions in 1919.

Later, in 1925, Kaluza collaborated with Oscar Klein (Swedish physicist, 1894-1977), to propose the use of higher dimensions as a way of building a unified theory of gravity and electromagnetism. Not much happened with this idea, but some years later, in 1943, Albert Einstein and Wolfgang Pauli (Austrian-Swiss physicist, 1900-1958) published a co-authored paper that cautiously approved of the Kaluza-Klein arguments.[146] Following that, in 1956, Pauli added, to a revision of

---

[145] Kurt Gödel, *Collected Works Volume III*, Eds. Feferman, Solomon (Oxford: Oxford University Press, 1995), 240.

[146] A. Einstein, and W. Pauli, "On the Non-existence of Regular Stationary Solutions of Relativistic Field Equations," *Annals of Mathematics*, Vol. 44, No. 2 (1943): 131.

his book, *The Theory of Relativity*, an extended supplemental note about the Kaluza-Klein solutions. In this supplement, he stated that although more work was need, this direction seemed on the right track.[147]

Unfortunately, the ideas put forward in the 1943 article, and the 1956 supplemental note, did not receive much attention. Nevertheless, these are examples where the concept of additional dimensionality was taken seriously, and in this case by two of the most prominent physicists of the 20th century.

As it turned out, all these ideas were essentially dormant until the 1980s, when String Theory, which later morphed into M-theory, became all the rage. Both theories relied on higher dimensions (10 for string theory and 11 for M-theory), although in these cases the extra dimensions are neatly compacted to the microscopic level to ensure that they remain out of sight. By *shrinking* the extra dimensions to the very small (Planck scale) these theories avoid both mathematical anomalies as well as not being required to explain why we do not experience the extra dimensions. However, after a period of lofty expectations, it became apparent that both String Theory and M-Theory were failing to live up to expectations. Subsequently, both have diminished in both luster and popularity.

---

[147] Pauli, 1981, 224.

The point is that interest in extra dimensions is not new, and it seems that interest has been growing over the last few years, possibly inspired by String Theory, such that dimensionality is again receiving serious attention. For example, in 1999, Lisa Randall (American physicist, 1962-) and Raman Sundrum (Indian-American physicist, 1964-) wrote a paper proposing that maybe all extra dimensions were not completely compacted,[148] and later Randall expanded on this notion in a book.[149] Another recent addition to the theme of extra dimensions is a book, published in 2006, by the physicist Paul Wesson (Canadian, 1949-2015).[150] These recent efforts favor extended extra dimensions rather than the rolled up compacted variety.

So, do not be surprised if somewhere in the not too far future someone comes up with something, but unlike the above examples, it will not be a fourth *spatial* dimension, but will reveal previously unknown hidden structure.

---

[148] Lisa Randall and Raman Sundrum, "An Alternative to Compactification." (1999) http://arxiv.org/pdf/hep-th/9906064.pdf.

[149] Lisa Randall, *Warped Passages* (New York, Ecco, 2005).

[150] Paul Wesson, *Five-Dimensional Physics* (London: World Scientific Publishing, 2006).

Final Comments

"The Red Queen shook her head. "You may call it 'nonsense' if you like," she said, "but I've heard nonsense, compared with which that would be as sensible as a dictionary!""[151]

"All that is known, all that is knowable, is set over in opposition against the absolute being of the object. The same ground, that assures us of the existence of things, shows them to be incomprehensible. Skepticism and mysticism are united in this one point. No matter how many new relations of 'phenomena' scientific experience may teach us, the real objects seem not so much revealed as rather the more deeply concealed by these relations."[152]

"Be it understood then that the writer is fully conscious of the indemonstrable nature of the hypothesis advanced, in so far as it is an hypothesis, and that he admits, in fact insists, that the value of such hypotheses lies not at all in themselves as ontological speculations about an Absolute, but in their success as expressions of the fundamental postulates and purposes of that source of all truth, Conscious Thought."[153]

Our common notions about space and time are the media for the ordering of the myriad affairs of humans. It is within these two concepts that we organize our lives such that we have come to

---

[151] Carroll 1871, 133. (I could not resist repeating this quote.)

[152] Cassirer 1953, 303.

[153] Royce 1882, 25.

believe that one is some universal container and the other is also a physical reality that provides sequence, continuity, and a certain direction. We intuitively need the container (space) because otherwise we would feel exposed and vulnerable, and we need time as a linear measure of order that flows only in one direction.

Although humans are primarily guided by common time, remnants of natural time still lurk in the background and continue to influence human behavior. For example, there still seems to be a certain allure and fascination with the full moon. Although most of us no longer associate the moon with lunacy, perhaps there is a thread of truth about the behavioral influence of a full moon.[154]

It is difficult to say which is easier to give up, common time or common space. The various arguments for the existence of physical space are different and yet similar to those for time. They differ in content but are similar in that they tend to take an anthropocentric view of the way things are in the universe. That is, there is a tendency to treat natural phenomenon as somehow connected to humans, as if human observation and

---

[154] Some years ago, a friend, who was an emergency center physician, told me that there seemed a definite connection between the full moon and activity at the emergency center. Perhaps various law enforcement agencies could corroborate this phenomenon.

interpretation of natural phenomena is a requirement for the existence of the natural world.

For example, we seem unable to avoid certain anthropic references and use gender specific names in references to both natural objects and human constructions. As another example, with the exception of English, other European languages take this tendency to the extreme by gendering nouns.

In the above discussions, the true nature of space and time were identified as convenient notions concerning movement, position, and change. That is, both time and space are mentally constructed relational notions that are so ingrained in our consciousness that we commonly regard them as some kind of stuff in the sense of having some kind of physical existence.

An unforeseen consequence of Minkowski's unity of space and time into the concept of space-time was that once united, they stand or fall together. However, even before this unity, time has never been able to stand completely on its own without the support of physical events. It is only through events (change, motion) that even natural time has ever had any meaning.

Even though the theories of relativity eliminated the notions of time and space as *some-thing* physical, I still see scientific and philosophical articles that refer to *physical time* and/or *physical space.* I can only assume that many people cannot imagine that without matter-energy content, there is only the Void, and that time only has meaning as a linear metric used to separate events, and that this measurement can only be performed by comparing one physical process (for example, A moving train) relative to another physical process (a clock). On the other hand, many people have trouble giving up the notion that space is a container of some kind and accepting that space only exists as relationships between objects.

Questions about space and time have their origins in the ancient past and it is reasonable to believe that these questions will continue to be discussed well into the future. There are always persons either unwilling or unable to put aside old ideas and adopt new ones. That is, realistically, there is no doubt that even with the insight into time and space that has been presented here, our notions of common time and space will not go away.

In any case, it was not the purpose here to get rid of common space and time. Both serve useful purposes within our everyday lives. It is only when contemplating deeper issues that the true nature of space and time must be considered such that common notions of space and time are relegated to their proper status. For example, when time and space are considered in the context of science or philosophy the true nature of time and space needs to be maintained.

That is, this identification of time and space as physical is not   a problem within everyday common usage and is useful for maintaining normalcy and the comfort factor that goes along with treating time and space as being some kind of stuff. It is only when we contemplate nature at its most fundamental level of our capabilities that the true nature of time and space must be considered.

*   *   *

Also discussed were various scientific theories of the recent past as well as those that are currently being formulated and how these current notions seem to lack the empirical content of past theories and the scientific method, so successful for the last few hundred years, seems to have been diluted. It seems that scientists today will too quickly accept a conjecture and even without any physical evidence proclaim that the conjecture is a fact. The prior discussion on the *Multiverse and Fine-Tuning Argument* is clearly such an example.

Perhaps it is worth repeating how the scientific revolutions of the early 20th century constitute paradigmatic changes of the first order. The science of the 20th century made possible the technology of the 21st century, and for better or worse, this technology has an increasing impact on all humanity and these changes have brought about a new way of thinking about reality and the natural world that is mechanistically inclined and self-focused.

When considering these paradigm changes, it is apparent that the most challenging is the new knowledge of matter and its behavior as revealed by quantum theory. With the development of quantum theory, science learned that the stuff of perceived creation was not what they had thought. As nicely stated by Ernst Cassirer (second quote at the beginning of this section), it seems apparent that as science probes ever deeper into the nature of substance, ultimate reality moves deeper into the shadows.

Additionally, we briefly took on the task of discussing the connections between perception and reality such that one always has influence on the other, and that although there are the physical objective aspects of perception, the subjective always has influence on perceptions and consequently reality is also modified with the subjective, either consciously or unconsciously.

Put another way, perception and reality were shown to be inextricably connected such that our reality is modified or supervened upon by subjective modifications to our perceptions that are used by many (most?) to buffer and isolate themselves from certain true realities of life and create the false notion that reality can be made different through acts of mostly symbolic posturing and gesturing such as *raising awareness,* which is usually just an empty phrase and a feel-good notion for the participant that is quickly forgotten and no real action takes place.

*   *   *

As well stated in the third quote at the start of this section, the hypotheses herein are unproven, and, for all our current knowledge (and future attempts?), the true essence of existence, most likely, will remain veiled behind the curtain. On the other hand, our knowledge of ourselves assures us that we exist in a way that is indisputable, and, by extension, we conclude that others exist and have thinking minds. Our thoughts are not mere illusions; they are why we know we exist.

Our knowledge is a fractional extraction from the inexhaustible content that lies beyond our current capability of perceiving. There is no final theory of everything. All theory is subject to revision or negation and the quest for such a theory of everything seems to have the characteristics of the alchemists' ancient and futile quest for the *Philosopher's Stone*. This is the legendary substance pursued by alchemists for centuries that was supposedly able, among other capabilities, to turn base metals into gold. Sometimes it was referred to as the *fifth element*.

What has been discussed here is not meant as more than a brief introduction and a mere hint at what lies both within and beyond the boundaries of our present knowledge and capabilities. The shadowy boundaries of today's knowledge will eventually be fully assimilated into the stock of human knowledge. As this happens, new boundaries, currently un-imagined, will appear and the cycles of challenge will continue. As the complexity of our thoughts increases new conceptions will arise and bring an understanding that becomes the starting point for the next advances in both knowledge and capability.

*   *   *

Current knowledge of the vastness of the universe and the unexpected strangeness of the micro-world require that we rethink some of our most strongly held beliefs. We need to remember that many of our beliefs originated thousands of years in the past, in a setting of a very small universe, where earth was the center of that universe.

Although modern cosmology has many questionable notions, enough of it is established to the point where rethinking our place in the universe is both desirable and necessary. For example, perhaps the effective universe for us is only a small portion of this cluster of stars that we call the Milky Way. At the other extreme of the *size* spectrum, the micro-world also requires that we rethink our ideas about the basis of fundamental substance.

*   *   *

The interaction of just two individuals is a reflection bound in the history of humankind. The very language we use is an expression of the history of our culture. Even though, on the surface, cultures have considerable variance, the underlying structure of all cultures reflects a universality, a wholeness that is interconnected so that the acts of humanity constitute a single drama.

Unfortunately, this universality does not mean that we are all the same or that we all want the same things in life. Just as there are diversities of language and culture, there are diverse worldviews and cultural objectives that can be incompatible with each other.

It is an uncomfortable truth that the current human condition, worldwide, makes it necessary that cultural entities be both open and closed. There must be a balance between being open and being closed that protects the integrity of a cultural tradition.

That is, where there is openness, there needs be assimilation into the hosting culture such that the aspects of the host culture that attracted in the first place are preserved and that openness is not used to undermine the current culture or to stand outside of the culture and create islands of disparate cultures. The infusion of diversity adds value only when it is integrated into the hosting culture.

*   *   *

The last comment will be about a word that has not yet been used. This is the word *hope*. Hope is the hidden strength of humankind and surpasses all other emotions. Hope is the emotion that drives us to continue with considered optimism that carries us through the darkest situations and most desperate hours.

"Knowing what is enough, you will not be humiliated.
Knowing where to stop, you will not be imperiled."[155]

---

[155] Lao Tzu 2005, 44.

Cited Works and General References

Aczel, Amir. *Entanglement*. New York: A Plume Book (Penguin), 2003.

Alexander, Samuel. "Space, Time, and Deity," *The Gifford Lectures at Glasgow*, 1920. https://archive.org/details/spacetime-anddeit00alexuoft>

Bergson, Henri. *Creative Evolution*. Translated by Arthur Mitchell. London: MacMillan and Co, 1911.

_____. *Matter and Memory*. London: George Allen & Unwin Ltd. Ruskin House, 1919.

Bertolam, Orfeu. "The Mystical Formula and the Mystery of Khronos," (2008). arXiv:qr-qc/0801.3994v1.

Bishop, R. and H. Atmanspacher. "Contextual Emergence in the Description of Properties," (2006) http://philsci-archive.pitt.edu/2934/

Bitbol, Michel, "Reflective Metaphysics: Understanding Quantum theory from a Kantian Standpoint," (2010) http://philpapers.org.

Bohm, David. *Wholeness and the Implicate Order*, (Routledge Classics, NY, NY, 1980).

_____. *The Undivided Universe*, (Routledge, New York, 1993).

Bruno, Giordano. *On the Infinite Universe and Worlds*, Venice, 1584. https://adoc.site/download/giordano-bruno-on-the-infinite-universe-and-worlds-a5b31f76133c56.

Capra, Fritof. *Tao of Physics*. Boston, MA: Shambala Publications 3rd Edition, 1991.

Carroll, Lewis. Through the Looking-Glass, and What Alice Found There (1871). New York: Barnes and Noble, 2012.

Cassirer, Ernst. *Substance and Function*. New York: Dover Publications, 1953.

Chalmers, David J. "Consciousness and its Place in Nature." S. Stich & T. Warfield, eds, *The Blackwell Guide to Philosophy of Mind* (Hoboken, NJ: Wiley- Blackwell, 2003).

Clifford, William K.: "On the Space Theory of Matter," *The World of Mathematics (Volume 1)*, Editor: James R. Newman, New York: Simon and Shuster 1956.

Darwin, Charles. *Origin of the Species* (1859). Complete works are available at darwin-online.org.uk

_____. *The Descent of Man* (1871). Complete works are available at darwin-online.org.uk

d'Espagnat, Bernard. *Veiled Reality* (Reading, MA: Addison-Wesley 1995).

Dewey, John. "Perception and Organic Action," The Journal of Philosophy, Psychology and Scientific Methods, Vol. 9, No. 24 (Nov.21, 1912), pp. 645-668

_____, *Art as Experience*. New York: Penguin Group, 2005.

Disney, Michael. "The Case Against Cosmology", arxiv: astro-ph/0009020v1, (1934) 2000.

Eddington, Arthur S. *The Nature of the Physical World* (Newcastle upon Tyne, UK: Cambridge Scholars Publishing, 1927), 286.

Einstein, A. and   Pauli, W. "On the Non-existence of Regular Stationary Solutions of Relativistic Field Equations," *Annals of Mathematics, Vol. 44 No. 2,* 1943.

_____, (1915): A. Einstein, *The Foundation of the Generalised Theory of Relativity* (1915): https://en.wikisource.org/

Ellis, George F. R. "Issues in the Philosophy of Cosmology. Section 2.8.1, Misconception 1," (2006) arXiv: astro-ph/0602280v2.

Epstein, P. F. "The Fine -Tuning Argument and the Requirement of Total Evidence", *Philosophy of Science Journal Volume 84 #4,* 2018.

Feynman, Richard. *QED: The Strange Story of Light and Magnetism.* Princeton, NJ, Princeton University Press, 1985.

_____, Feynman, Richard. *The Feynman Lectures on Physics: Volume* I. Reading MA, Addison-Wesley, 1989.

Gilder, George. *Life after Google*, Washington DC, Regnery Gateway, 2018.

Gödel, Kurt: *Collected Works Volume III.* Oxford: Oxford University Press, Eds. Feferman, Solomon, 1995.

Grandpierre, Attila. "The Dynamics of Time and Timelessness," In the Nature of Time: Geometry, Physics and Perception. *NATO Science Series II. Mathematics, Physics and Chemistry -*

Vol. 95, Netherlands, Kluwer Academic Publishers, 2003.

Greene, Brian. *The Elegant Universe*. New York: W. W. Norton & Company, 2003.

Hartnett, J. G. (2011): "Is the Universe Really Expanding?" https://arxiv.org/abs/1107.2485

Hawking, S. A Brief History of Time: From the Big Bang to Black Holes. New York, NY: Bantam Books, 1988

_____. "Gödel and the End of Physics," (http://www.hawking.org.uk/godel-and-the-end-of-physics.html).

Herodotus. *The Landmark Herodotus: The Histories*, Ed. Robert Strassler, New York: Pantheon Books, 2007.

Holland, John H. *Emergence: From Chaos to Order*. Reading, MA: Perseus, 1998.

Hubble, Edwin. "Effects of Red Shifts on the Distribution of Nebula," *American Astronomical Society – Provided by NASA Astrophysics Data Systems*. (1936) http://adsabs.harvard.edu/cgi-bin/basic_connect?qsearch=edwin+hubble&version=1

Huxley, Aldus. *The Perennial Philosophy*. New York-London: Harper Modern Classics, 2009.

James, William. *Principles of Psychology Vol. 2*. Cambridge, MA: Harvard Univ. Press, 1910.

_____. *Principles of Psychology the Briefer Course*. Cambridge MA: Harvard University Press, 1923.

_____. *The Varieties of Religious Experience*. New York NY: Longmans, Green, And Co, 1917.

Janis, I. L. Groupthink: Psychological Studies of Policy Decisions and Fiascoes. Boston (MA, USA): Houghton Mifflin, 1972.

Jung, Carl. *Synchronicity: An Acausal Connecting Principle*. Princeton and Oxford: Princeton University Press, 1960.

Kragh, Helge. "The Controversial Universe: A Historical Perspective on the Scientific Status of Cosmology". (2007) https://d-nb.info/101824364X/34.

Kroupa, Pavel. "The dark matter crisis: falsification of the current standard model of cosmology."(2012) https://arxiv.org/abs/1204.2546v2.

Lao Tzu, *Tao Te Ching* (New York: Barnes and Noble Classics, 2005).

Leighton, J, A. "Perception and Physical Reality," *The Philosophical Review, Vol. 19, No. 1 (Jan. 1910), pp. 1-21*. Duke University Press on behalf of Philosophical Review Stable. (1910) http://www.jstor.org/stable/2177636.

Linde, Andre. "The Uniformity and Uniqueness of the Universe," *This Idea Must Die*: Ed. John Brockman, (Harper Perennial, New York, 2015).

McTaggart, John Ellis. "The Unreality of Time," Mind: *A Quarterly Review of Psychology and Philosophy* 17: pages 456-473. (1908) http://www.ditext.com/mctaggart/time.html.

Meister Eckhart. *Meister Eckhart, A Modern Translation*. Trans: R. B. Blakney (New York: Harper & Brothers, 1941).

Merleau-Ponty, Maurice. *The Primacy of Perception*. Evanston, IL: Northwestern University Press, 1964.

Milne, Edward Arthur. *Modern Cosmology & the Christian Idea of God*. Oxford: Oxford Press, 1951.

Nicholas of Cusa, *De Docta Ignorantia II (Learned Ignorance II)*, Chapter Three (1440). http://www.jasper-hopkins.info/DI-Intro12-2000.pdf http://www.jasper-hopkins.info/DI-Intro12-2000.pdf

_____, *Vision of God*. https://archive.org/details/TheVisionOfGodByNicholasOfCusa-OphthalmosAploisOrTheSingleEye-/page/n0

Oppenheimer, J. R. "The Oppenheimer Years," *Los Alamos Science Winter/Spring* (1983) http://la-science.lanl.gov/lascience07.shtml.

Pauli, Wolfgang. *The Theory of Relativity*. New York: Dover Publications Inc, 1981.

_____. *Writings on Physics and Philosophy*. Eds. C. Eng and K. von Meyenn, Trans: R. Schlapp, 1994.

Penrose, Roger. *The Road to Reality*, Ny, NY, Alfred Knopf 2006.

Poincaré, Henri. *Science and Method*. London: Thomas Nelson and Sons (Trans. Francis Maitland, 1918 edition).

_____. "The Measure of Time," *The Foundations of Science (The Value of Science)*. New York: Science Press, pp. 222-234. (1913) archive.org/details/foundationsscie01poingoo g.

_____. *Science and Hypothesis*. New York, NY: Dover Publications Inc., 1952 (Originally published 1902).

Popper, Karl. Conjectures and Refutations: The Growth of Knowledge. New York: Harper and Roe, 1963.

Pratt, James. "Realism and Perception," *The Journal of Philosophy, Psychology and Scientific Methods*, Vol. 16, No. 22, pp. 596-603, (1919) http://www.jstor.org/stable/2940132.

Pseudo-Dionysius the Areopagite, *Mystic Theology*. T. Davidson translation, *Journal of Speculative Philosophy*, 1893, vol. xxii.

Randall, Lisa and Sundrum, Raman. "An Alternative to Compactification." (1999) http://arxiv.org/pdf/hep-th/9906064.pdf

_____. *Warped Passages*. New York: Ecco, 2005.

Reichenbach, Hans. *The Direction of Time*. New York: NY, Dover Publications Inc., 1984.

Riemann, Bernhard: "On the Hypotheses which lie at the Bases of Geometry" in *God Created the Integers*, Philadelphia – London: Running Press, Editor: S. Hawking, 2007.

Royce, Josiah (1882): "Mind and Reality," *MIND*, (1882) https://archive.org/details/mindreality00roycrich

_____. *The Religious Aspects of Philosophy*. Boston and New York: Houghton Mifflin Company, 1885.

Russell, Bertrand. *The Philosophy of Logical Atomism*. London and New York: Routledge Classics, 2010.

Sellars, R W: "Panpsychism or Evolutionary Materialism," *Philosophy of Science*, vol 27, number 4, 1960.

Smolin, Lee. "The Big Bang was the First Moment of Time," *This Idea Must Die*: Edited by John Brockman, (Harper Perennial, New York, 2015).

_____. *The Life of the Cosmos*. N Y, NY: Oxford UP, 1997.

Stapp, H. P. *Mind, Matter, and Quantum theory*. Berlin Heidelberg: Springer-Verlag, 2009.

St. Augustine of Hippo. *Confessions of St Augustine*. New York: Modern Library, 1999.

Wang, Hao. *A Logical Journey*. Cambridge MA: MIT Press, 1996.

Wesson, Paul. *Five-Dimensional Physics*. London: World Scientific Publishing, 2006.

Weyl, Hermann. *Philosophy of Mathematics and Natural Science*. Princeton, NJ: Princeton University Press, 1949.

Whitehead, Alfred North. *The Concept of Nature*. Cambridge: Cambridge Press, 1920.

Wilber, Ken. *Quantum Questions*. Boston MA: Shambhala Publications, 2001.

Yourgau, Palle. *A World Without Time*. Cambridge, MA: Basic Books, 2005.

Zukav, Gary. *The Dancing Wu Li Masters*. New York: Perennial Classics, 2001.

Index